RESEARCH AND DEVELOPMENT DATA NEEDS:

PROCEEDINGS OF A WORKSHOP

Bronwyn H. Hall and Stephen A. Merrill, Editors

Board on Science, Technology, and Economic Policy

Policy and Global Affairs

NATIONAL RESEARCH COUNCIL
OF THE NATIONAL ACADEMIES

THE NATIONAL ACADEMIES PRESS
Washington, D.C.
www.nap.edu

THE NATIONAL ACADEMIES PRESS   500 Fifth Street, N.W.   Washington, DC   20001

NOTICE: The project that is the subject of this report was approved by the Governing Board of the National Research Council, whose members are drawn from the councils of the National Academy of Sciences, the National Academy of Engineering, and the Institute of Medicine. The members of the committee responsible for the report were chosen for their special competencies and with regard for appropriate balance.

This study was supported by Contract No. CNST-1-02-03-B between the National Academy of Sciences and the National Science Foundation. Any opinions, findings, conclusions, or recommendations expressed in this publication are those of the author(s) and do not necessarily reflect the views of the agency that provided support for the project.

A PDF is available at www.nap.edu.

Limited copies are available from:

Board on Science, Technology, and Economic Policy
National Research Council
500 Fifth Street, N.W.
Washington, D.C. 20001
Phone: 202-334-2200
Fax: 202-334-1505

Copyright 2005 by the National Academy of Sciences. All rights reserved.

THE NATIONAL ACADEMIES
Advisers to the Nation on Science, Engineering, and Medicine

The **National Academy of Sciences** is a private, nonprofit, self-perpetuating society of distinguished scholars engaged in scientific and engineering research, dedicated to the furtherance of science and technology and to their use for the general welfare. Upon the authority of the charter granted to it by the Congress in 1863, the Academy has a mandate that requires it to advise the federal government on scientific and technical matters. Dr. Bruce M. Alberts is president of the National Academy of Sciences.

The **National Academy of Engineering** was established in 1964, under the charter of the National Academy of Sciences, as a parallel organization of outstanding engineers. It is autonomous in its administration and in the selection of its members, sharing with the National Academy of Sciences the responsibility for advising the federal government. The National Academy of Engineering also sponsors engineering programs aimed at meeting national needs, encourages education and research, and recognizes the superior achievements of engineers. Dr. Wm. A. Wulf is president of the National Academy of Engineering.

The **Institute of Medicine** was established in 1970 by the National Academy of Sciences to secure the services of eminent members of appropriate professions in the examination of policy matters pertaining to the health of the public. The Institute acts under the responsibility given to the National Academy of Sciences by its congressional charter to be an adviser to the federal government and, upon its own initiative, to identify issues of medical care, research, and education. Dr. Harvey V. Fineberg is president of the Institute of Medicine.

The **National Research Council** was organized by the National Academy of Sciences in 1916 to associate the broad community of science and technology with the Academy's purposes of furthering knowledge and advising the federal government. Functioning in accordance with general policies determined by the Academy, the Council has become the principal operating agency of both the National Academy of Sciences and the National Academy of Engineering in providing services to the government, the public, and the scientific and engineering communities. The Council is administered jointly by both Academies and the Institute of Medicine. Dr. Bruce M. Alberts and Dr. Wm. A. Wulf are chair and vice chair, respectively, of the National Research Council.

www.national-academies.org

Planning Committee for Workshop to Review Research and Development Statistics at the National Science Foundation

BRONWYN H. HALL (Chair), Department of Economics, University of California, Berkeley, CA
WILLIAM G. BARRON, JR., Department of International Affairs, Woodrow Wilson School, Princeton University, Princeton, NJ
LAWRENCE D. BROWN, Department of Statistics, Wharton School, University of Pennsylvania, Philadelphia, PA
ROBERT H. MCGUCKIN, The Conference Board, New York, NY
DONALD SIEGEL, Department of Economics, Rensselaer Polytechnic Institute, Troy, NY
GREGORY TASSEY, National Institute of Standards and Technology, Gaithersburg, MD
RICHARD TURMAN, Association of American Universities, Washington, DC
PATRICK WINDHAM, Windham Consulting, Atherton, CA
ANDREW W. WYCKOFF, Organization for Economic Co-operation and Development, Paris, France

STEPHEN A. MERRILL, *Study Director*
CRAIG SCHULTZ, *Senior Research Associate*

Board on Science, Technology, and Economic Policy

CHAIRMAN
Dale Jorgenson
Samuel W. Morris University Professor
Harvard University

VICE CHAIRMAN
Bill Spencer
Chairman Emeritus
International SEMATECH

M. Kathy Behrens
Managing Director for Medical Technology
Robertson Stephens Investment Management

Kenneth Flamm
Professor and Dean Rusk Chair in International Affairs
LBJ School of Public Affairs
University of Texas-Austin

Bronwyn Hall
Professor of Economics
University of California, Berkeley

James Heckman
Henry Schultz Distinguished Service Professor of Economics
University of Chicago

David Morgenthaler
Founding Partner
Morgenthaler Ventures

Mark B. Myers
Visiting Executive Professor of Management
The Wharton School
University of Pennsylvania

Roger Noll
Morris M. Doyle Centennial Professor of Economics
Director, Public Policy Program
Stanford University

Edward E. Penhoet
Director, Science and Higher Education Programs
Gordon and Betty Moore Foundation

William J. Raduchel
Chairman and CEO
Ruckus Network

Jack Schuler
Chairman of the Board of Directors
Ventana Medical Systems, Inc.

STAFF

Stephen A. Merrill
Executive Director

Charles Wessner
Deputy Director

Sujai Shivakumar
Program Officer

Craig Schultz
Senior Research Associate

McAlister Clabaugh
Program Associate

David Dierksheide
Program Associate

PREFACE

This report contains the proceedings of a one-day workshop organized by the National Research Council's Board on Science, Technology, and Economic Policy (STEP), in conjunction with a study by a panel of the NRC Committee on National Statistics (CNSTAT). This combined activity was commissioned by the Science Resources Statistics Division (SRS) of the National Science Foundation (NSF) to recommend improvements in the Foundation's portfolio of surveys of research and development spending by the federal government, state governments, private industry, the nation's universities and colleges, and other nonprofit institutions

The purpose of the workshop was to inform the CNSTAT panel and the Foundation about the adequacy of national and regional R&D data from the perspective of evaluating economic performance and formulating public policies to enhance it. Not designed to review the entire portfolio of R&D surveys, the workshop examined several areas where specific concerns have been expressed Thus, after first considering some users' data needs and recent changes in the surveys, the agenda primarily focused on the distribution of R&D spending among research fields and industry sectors and across geographic regions, the nature of R&D activity in the service sector, and the extent of cross-national and collaborative R&D investment by large firms. The CNSTAT panel's recommendations on the broader topic of NSF's portfolio of R&D data surveys are contained in the report, *Measuring Research and Development Expenditures in the U.S. Economy* (National Research Council, 2004).

The workshop was the third in a series of meetings convened by the STEP Board to assess national indicators and data on innovation processes encompassing but not limited to formal research and development. The first meeting, sponsored by the National Science Foundation, was held in February 1997 and was intended to generate suggestions for improving the measurement of industrial research and innovation activity. It resulted in the report, *Industrial Research and Innovation Indicators* (National Research Council, 1997). The second meeting, sponsored by the National Aeronautics and Space Administration, National Institutes of Health, and Department of Energy, was held in November 1999 to consider how data on scientists, engineers, and other professionals--their training, employment and mobility, structure of work and affiliations, and productivity–could be used to illuminate trends in industrial innovation. It resulted in the report, *Using Human Resource Data to Track Innovation* (National Research Council, 2002).

The STEP program has not been concerned solely with private sector innovation. It has also examined the United States' public and private investment in research to generate new scientific and technical knowledge and the relationship of that research investment to the training of scientists and engineers at the post-graduate level. In a 2001 report, *Trends in Federal Support of Research and Graduate Education* (National Research Council, 2001) a STEP committee examined the 1993-2000 trends in research spending in 23 disciplines and the corresponding trends in graduate enrollment (by source of financial support) in those disciplines, primarily to determine the magnitude of the shift in spending and enrollments away from many physical science and engineering fields in the aftermath of the Cold War. This analysis relied primarily on NSF's Survey of Federal Funds for Research and Development but it also considered data from the Survey of Research and Development Expenditures at Universities and

Colleges, the Survey of Industrial Research and Development, the one-time (FY 1995) Survey of State Research and Development Expenditures, and the Survey of R&D Expenditures by Nonprofit Institutions as well as the Foundation's surveys of science and engineering personnel.

Research and development are embedded in two broader intersecting processes, both of which are important to economic growth, productivity advances, and employment creation as well as to noneconomic aspects of national welfare–physical security, environmental preservation, intellectual understanding, and cultural creativity. The first process is technological innovation, which refers to the invention, commercialization, and diffusion of new products, processes, and services. Innovation is an important determinant of aggregate productivity and economic growth, countries' comparative advantages, and firms' competitiveness vis-à-vis their rivals. The second process is creation and diffusion of new knowledge, which may or may not have near-term market-based applications but which represents a pool of information from which many different kinds of users draw. To understand these processes, conceptual and data needs extend well beyond current national R&D statistics, whose focus is narrowly on how much money is spent by whom on the performance of what types of research and development in which institutions.

Consider first industrial innovation. Although theoretical understanding of the process is not far advanced, it is possible to elaborate a simple conceptual framework that encompasses the aspects that it is important to measure if we are to have a full picture of the status of national innovation. Four broad categories of innovation information are (1) innovation *inputs* such as the level and direction of the research effort but also activities that occur outside of formal R&D laboratories, for example, changes in management practices or on the manufacturing line; (2) determinants of innovative activity such as the means and effectiveness of appropriating returns from innovation, market structure and conditions, the nature and extent of knowledge flows among firms and other institutions, technological opportunities, etc., (3) measures of innovative *output* – newly commercialized products, processes, and services and their characteristics and diffusion; and (4) *outcomes* or the impact of innovation on firms, regions, workers, and the economy as a whole. On some of these important variables hardly any data are currently collected on a national basis and regular schedule for major sectors of the economy; on other variables information is fragmentary.

- Systematically collected data are generally lacking on new product, process, and service introductions and their use and diffusion, although some intermediate outputs, including patents and their citations, are well documented.
- Apart from formal R&D, we have little idea how much innovative activity goes unmeasured.
- There are large gaps in data on the many channels (e.g., meetings, personal communications, technical publications, patents, etc.) through which knowledge moves within firms, between customers and suppliers, and across competing firms, along with other determinants of innovative activity.
- Macro-level data on GDP and productivity growth are readily available but micro-level data on firm performance and worker characteristics likely to be influenced by innovation are not.

These deficiencies are a principal rationale for the efforts – extensive, publicly supported, and internationally coordinated in Europe, highly detailed in Canada and Australia, but much more modest in the United States – to mount "innovation surveys," independent of R&D reporting, to fill some of these gaps. It was part of the CNSTAT panel's mandate to evaluate these efforts and make recommendations to NSF on conducting research and surveys on innovation (NRC, 2004, Chapter 4).

For scientific and engineering knowledge creation the principal data sources are surveys of the training and deployment of scientists and engineers, publications of science and engineering research results, and formal and informal channels apart from publications by which new knowledge is communicated – licenses, meetings, consulting arrangements, personal communications, etc. The quality and robustness of the data collected and reported by NSF are greatest for the first two categories – training and publications -- at least for the individuals and journals with recognized standing in science and engineering (e.g., individuals with degrees from accredited institutions and journals with professional society affiliations). Data are more limited for contractual relationships such as licenses and consulting and poorest for informal channels of communication.

Research and development merit close attention because they are critical inputs to both innovation and knowledge creation and diffusion. Moreover, because of budgeting, accounting, and tax conventions, data on R&D expenditures are easier to collect than are data directly measuring innovation and knowledge creation. For these reasons, R&D expenditures are often treated as surrogates for knowledge creation and innovation effort. In addition, some administrative data[1] exist on industrial R&D collaborations among companies and between companies and federal laboratories.

Notwithstanding the extensive portfolio of R&D expenditure data, from time to time several concerns about their adequacy and comprehensiveness have been articulated. Some of the concerns are longstanding while others derive from recent perceptions that the process of innovation and characteristics of R&D are changing. These concerns motivated the Foundation's request. They relate to

- the comprehensiveness of data on some payers and performers of R&D (e.g., state and local government and nonprofit institutions that are surveyed irregularly);
- the classification of R&D by type (basic research, applied research and development), by industry, sector or activity, and by field of science and engineering as well as consistency of taxonomies across surveys;
- the geographical location of R&D, both internationally and by locality within the United States;
- Characteristics of R&D performance (locus within the institution or firm, time horizon, etc.);
- inconsistencies in what are ostensibly the same data reported by different sources (e.g., differences between federal spending in universities as reported by federal agencies and performing institutions);
- timeliness of R&D data availability; and
- Survey methodology (e.g., sample construction, imputation, etc.).

This workshop addressed selected issues in the first four categories.

Reflected in all three of the workshops organized under STEP's auspices is a perception that these and other issues of R&D data collection have become more important because of the likelihood that the pattern, determinants, and effects of innovation are changing, sometimes remarkably quickly and radically and with potentially significant consequences for the nation's economic performance. These perceived changes include shifts in the sectoral, industry, and

---

[1] Administrative data are statistics gathered by government in the course of performing some function other than information gathering per se -- e.g., regulation (SEC filings), taxation, subsidization, etc.

technology distribution of innovative activity, shifts in the time horizon of innovative effort and investment, changes in the organizational structure of innovative activity, and changes in the location of activity both within the United States and globally.

Officially reported national data should not be expected to capture all of the R&D trends of interest, let alone tell us all that we want to know about knowledge creation and innovation processes. Throughout the workshop there were descriptions of private data collection efforts of great utility, some of them of them repeated periodically, others limited to a single project. These are essential complements to NSF's continuing survey time series and often necessary to illuminate discontinuities and emerging changes.

The workshop was planned by a diverse committee including experts in statistics, economics, policy analysis, and policymaking. Their institutional experience is also wide-ranging. Two are veterans of a federal statistical agency, two of congressional staffs, one of a federal R&D agency, and one of the White House staff. State government, international business, and international organizations were also represented, and two panel members have spent their careers primarily in academic research.

The meeting was organized in seven parts. In the first session several government officials described their R&D data needs for national economic accounting and program planning and evaluation purposes. An official of the National Science Foundation's Office of Science Resources Statistics followed with a detailed description of recent changes in the R&D survey data portfolio – some of them responsive to new needs and criticisms, others made for reasons of practicality and reliability. There were two discussions of the composition of R&D, one focused on the field distribution of public sector basic and applied research and the other focused on the distribution of business R&D activity. The adequacy of data on R&D collaborations across firms was discussed. Finally, there were two sessions on locational issues – one addressing cross-national R&D investment and the other examining the domestic distribution of R&D activity.

The views expressed in the STEP workshop were, of course, those of the individual speakers and do not necessarily represent positions of the National Academies. Nevertheless, information and insights from the presentations were incorporated in the CNSTAT panel report along with information from its own two-day workshop and many other sources.

Bronwyn H. Hall
Stephen A. Merrill

ACKNOWLEDGMENTS

This volume has been reviewed in draft form by individuals chosen for their technical expertise, in accordance with procedures approved by the NRC's Report Review Committee. The purpose of this independent review is to provide candid and critical comments that will assist the institution in making its published report as sound as possible and to ensure that the report meets institutional standards for quality. The review comments and draft manuscript remain confidential to protect the integrity of the process.

We wish to thank the following individuals for their review of the preface: Lawrence Brown, University of Pennsylvania; Wesley Cohen, Duke University; and Thomas Plewes, consultant. Although the reviewers provided constructive comments and suggestions, they were not asked to endorse the content of the preface. Responsibility for the final content of this proceedings rests with the editors and the individual workshop presenters.

We also thank Marisa Gerstein, Tanya Lee, and Lee Pollack of the staff of the NRC's Committee on National Statistics for their initial editing of the workshop transcript.

CONTENTS

INTRODUCTION.. 1

USERS AND USES OF R&D DATA.. 1

RECENT DEVELOPMENTS IN NSF's R&D DATA

      PORTFOLOIO.. 11

COMPOSITION OF PUBLIC SECTOR RESEARCH... 13

COMPOSITION OF INDUSTRIAL R&D.. 17

R&D COLLABORATIONS... 25

LOCAL R&D ACTIVITY... 31

INTERNATIONAL R&D ACTIVITY.. 33

CONCLUDING OBSERVATIONS... 41

APPENDIX A: WORKSHOP AGENDA... 46

APPENDIX B: LIST OF PARTICIPANTS.. 49

INTRODUCTION

BRONWYN HALL: This workshop, being conducted by the Board on Science, Technology, and Economic Policy, is part of a Committee on National Statistics study of the national portfolio of research and development statistics – a study commissioned by the National Science Foundation. We use the term R&D but actually we mean research, development, and innovation broadly defined. Its mandate is to examine the uses and types of data that are currently being collected and to determine how they might be improved to reflect what is generally considered a rapidly changing R&D enterprise. More specifically, we are exploring how to improve measures of the composition structure, performers, and the geographic location of R&D sponsored or paid for by industry and the federal government, which account for the majority of R&D activity.

The first panel will talk about who uses the data and what uses they make of them. Then we will hear from the National Science Foundation office responsible for the R&D surveys, the Office of Science Resources Statistics, what changes have been made in them recently. Then we will turn to some of the larger problems of classification, first the division of R&D into basic research, applied research, and development and the subdivision of basic and applied research into research fields and, secondly, the composition of industrial research and development. This is a particularly important issue in the service sector, where R&D and growing most rapidly.

In the afternoon we will consider how to capture the increase in research and development collaborations in our statistics. Finally, the geographic location of R&D has macro and micro aspects. R&D is done in other countries by U.S.-headquartered firms and R&D is performed in the United States by firms from other countries. How well are these patterns captured? Second, many, including members of Congress, would like a more detailed breakdown of R&D carried out in individual states and economic regions.

USERS AND USES OF R&D DATA

BARBARA FRAUMENI: : Today I am going to discuss, from the perspective of the Bureau of Economic Analysis, why we want R&D data and what kind of data we would like to have.

My coauthor Sumiye Okubo and I recently wrote a paper about the contribution of R&D to economic growth. We integrated R&D into the system of National Accounts by putting it on the product side and the income side, in good national income accounting practice. The major thing we did was treat R&D as investment. And we concluded that over the 40-year period from 1961 to 2000, R&D accounted for approximately 10 percent of growth in GDP. In other words, if GDP grew at 3 percent a year on average during that time period, R&D accounted for 0.3 of that growth rate. This roughly corresponds to the number that others have obtained for the contribution of computer hardware to growth in GDP during the wonder years of the second half of the 1990s, so it is an impressive number.

Second, treating R&D as investment raises the savings rate quite appreciably, from 19 to 21 percent. Capitalizing R&D does very little to change the rate of growth of GDP. So our conclusion is that we are not really misstating economic growth, but by incorporating R&D in the accounts we do have a better sense of what is bringing about growth. In short, we want R&D data because it is a very important source of economic growth and we need to understand the process better.

When the System of National Accounts (SNA), which is used by almost every country in the world, was last revised we decided not to capitalize R&D. However, there is an international meeting coming up shortly in the Netherlands, and it may decide that R&D should be capitalized. Once that happens there will be a flurry of activity on R&D. As far as I know the United States has the only complete set of satellite accounts incorporating R&D. Israel has a partial set of accounts, and France has a set of accounts that encompasses information technology but doesn't capture other parts of R&D. If the decision is made to capitalize R&D others will seek our advice about how to do it. At the meeting in the Netherlands participants will consider how R&D might be capitalized in the System of National Accounts and the relationship between SNA and the *Frascati Manual,* the OECD's guide to the collection of R&D data.

Let me turn to our R&D data wish list. My coauthor and I spent very little time looking at the quality of the data; we took them as given. But as we proceeded we did observe some things, and we did reach some judgments about what it would be desirable to have.

First, most people really want to know what kind of R&D performed in what industries contributed most to the increase in economic growth, but we could not begin to answer that. Missing from our satellite account is information on industries and the intermediate as well as capital and labor inputs to the production process, including all the inputs that enter into an R&D activity. We were working on a performing basis but we knew nothing about the industry composition because we can't get enough data from NSF to perform this sort of analysis. So our foremost need is for industry detail, particularly information on services.

We also were concerned about things like outsourcing. Who does R&D for IBM? Is it IBM in its manufacturing area? Is it another company outside manufacturing? Is it in the services? IBM was doing R&D somewhere, but where was it? So we do need information on services, not just manufacturing. If you look at the R&D data for manufacturing, the industries that you want to look at most carefully have missing entries because of reporting problems because there are only a few companies in those cells.

Secondly, inputs and outputs in the national accounts work on an establishment basis. A company like IBM has a wide variety of establishments with a large percentage of the activity in services rather than computer hardware. So if you really want to know who is doing what to get a true picture of R&D, you want an establishment basis so you can clearly identify companies that are doing a significant amount of R&D, but have a large proportion of other activities.

We want even more information about industrial activity. There is a big difference between basic, applied, and development R&D -- a difference in the time before it reaches the economy. In recent years there has been a shift toward development, in large part because there has been a marked increase in business performance of R&D relative to the government. As that shift proceeds, there is also a shift away from basic towards applied research.

Later today some of my colleagues are going to come here to speak about the international aspect of R&D. We only have a few years of data and only for certain types of companies. We would like more of this because a lot of the R&D could be performed abroad. Also we would like to know what sort of R&D is imported and what are the spillovers of R&D. International boundaries do not mean a great deal in this context.

A continuous time series is important. We know that there has been at least one series break as result of a substantial change in the survey. If I recall correctly it had to do with the number of companies that were surveyed, and it occurred sometime in the latter part of the 1980s or the early part of the 1990s, periods that people are very interested in. We were told by NSF that their attempts to bridge this problem were unsuccessful. I would like to see that bridge completed because we really do want to have a continuous time series to enable fair comparisons over time.

Next, the issue of industry classification system, NAICS versus SIC. We are in the throes of converting to a new industry classification system, and as we do this, in many cases we are dropping the historical time series because we are finding it difficult to link industry definitions and trace them back through time. It is important that we be able to do this.

The spillover question. A lot of R&D tends to be concentrated in certain areas. I heard a paper by a Ph.D. candidate the other day who was trying to determine to what extent physical distance mattered; he was looking at R&D by state. Well, we have little information on the geography of R&D.

What is the relationship between R&D performers? For that matter, what is the relationship between funders and performers? We go to great effort in our satellite account to get everything on a performance basis. Nevertheless, it can matter who the funder is and there may be associations between performers.

Micro data. In my dreams I want to be able to go to NSF and use data that cannot be put in a publication because of confidentiality concerns, much in the same way that you can go to Census Research Centers and access the micro data under strictly controlled conditions. This would be a way of getting around the problem of missing data because of industry sensitivity. It will take time to bring something like this to pass.

Finally, bring on another Mansfield! Ed Mansfield did several studies a number of years ago to ascertain the benefits of R&D. It is important in our paper in the context of who is receiving the spillover benefits of R&D. I would love to have another Mansfield emerge to give us more information on the nature and types of benefits arising from R&D.

The next speaker is Andrew Wyckoff from the Organization for Economic Co-operation and Development.

ANDREW WYCKOFF: Our presentation is in two parts, the first part by me and a second by my colleague in the Directorate for Science, Technology, and Industry, Dominique Guellec, who heads the unit responsible for the *Frascati Manual* and some of the methodologies that underpin what we are discussing today.

A word about the OECD and what we do. We are a 30-member organization of industrialized countries headquartered in Paris. We may be best known for trying to produce reasonably harmonized data in various areas, including R&D. We deal with the details of methodology and try to arrive at some agreement about how to proceed. The heavy lifting is done by the member country delegates to the committee of National Experts in Science Technology Indicators (NESTI).

From where we sit we see a number of different types of users and uses of our data, but by far the most popular is simply basic comparisons of R&D levels or R&D intensities, normalized by the size of the population or the size of the economy, the GDP. That is by and far the most common use of the data, but we also see comparisons of industrial structure, particularly targeting on what are thought of as strategic technologies. In addition, there is a lot of interest in how the 30 countries, which are roughly similar in stage of economic development, differ in both the funding and the performance of R&D. Equally important are the human resources associated with R&D, but I will not talk as much about that as about R&D expenditures.

The topic du jour of the OECD on R&D has got to be the targeting of various R&D

intensities by some of our member countries; most notably, the European Union has set a target of 3 percent R&D intensity they hope to be approaching by 2010. But you can see a lot of other countries are doing this type of targeting activity, some of them making explicit references to our data, which makes us a bit nervous because there is imprecision in this data that we will talk about today.

Especially as the EU becomes more integrated in its policies, the most relevant comparison for the EU economy is to that of the United States. Not only is the United States the benchmark in terms of R&D intensity, but also in terms of the performance of R&D. There is an effort in Europe to make their funding structure look similar to that of the U.S., particularly by emphasizing the business funding of R&D.

Let me shift now from the policy arena to some of the data needs that we see. I suspect we are going to hear these repeatedly today. One is the problem of common definitions of R&D, particularly with respect to borderline activities, such as software development, where some of the activity is R&D, and some of it is not. And military equipment prototypes are a big item in the United States and handful of other countries that invest heavily in defense R&D.

There are other trouble spots. Industrial coverage may differ across member countries, particularly with respect to how thoroughly they survey the service sector. We do not know as much as we would like about countries' treatment of multinationals. We are looking carefully at that issue. There are also important institutional differences. An illustration is the very large and important health sector. If you use government budget appropriations and you break it down by objective, the U.S. appears to be an outlier with a very large bias towards health. But in fact part of this represents just cultural and institutional differences across the OECD. When you aggregate some of the categories and bring them together so they are more comparable, the U.S. is actually less an outlier than it originally appeared. This is the type of work that we have to do at the OECD.

DOMINIQUE GUELLEC: What is the OECD doing to satisfy some of the needs that are on this workshop's agenda? As we have said we in the OECD are in the ambiguous position of being on the demand side as user, but also we are part of the supply chain in helping countries to set statistical and data collection standards.

So let us first look at the field of science classifications, which is a disaggregation of R&D expenditure by major fields such as physics and biotechnology and so on. There is a clear need for such data at the country level, since governments want to check the consistency of their science policies with the needs articulated by their industries. If a country has a very strong chemical industry while R&D funding is going primarily to physics, you may have an inconsistency that the government may be willing to correct. Also at the international level we may want to compare countries' respective contributions to particular fields such as biotechnology and others. Over the past several years NESTI has arrived at an agreed-upon definition of what biotechnology is, or really two definitions. On this basis international data will become available in the next year or two. It is already available from certain countries.

More broadly, we have begun to develop a new classification of scientific fields. The one we have in the *Frascati Manual* now dates back to the 1960s, and it came from UNESCO. It was good for its time but science is changing. We have sent a questionnaire to countries asking about their thinking on the proper classification of science, and we have begun to receive responses.

Second, the breakdown of R&D by type. Basic research / applied research / development is the principal breakdown and it is useful   We have the notion that different types of R&D will affect economic growth over different time scales; this is important to know for monitoring policy. A well-known difficulty, however, is that the borders of these categories may be different across countries and across industries. I believe that a more careful reading of the definitions in the *Frascati Manual*, both by respondents to surveys and by the statisticians who carry them out, would help clarify these concepts. For example, there is a common notion that basic research has no application. That is not what the *Frascati Manual* actually says. It considers basic research without application as a particular type of basic research that we call non-oriented basic research. So it is not always our methodologies that are lagging but sometimes our understanding of these methodologies. At the moment approximately one-half the OECD member countries collect basic research data.

Next, R&D by industry is needed for understanding the dynamics of productivity. It is rather difficult to collect good data in this respect. There are multi-industry groups and specialized R&D companies, and both phenomena are increasing to the point that they really represent their own industry. And this R&D industry is, of course, serving other industries and its output is affecting productivity in these other industries, making it difficult to know what the input/output relationships among them are.

In response, we at the OECD are emphasizing the industry of use. Many software firms, for example, are developing products for the automobile industry. Their work, according to *Frascati*, should be classified as serving the automobile industry not the software industry. That makes sense, of course, if you are interested in productivity effects.

Actually, we have been collecting and harmonizing industry data for some time and now we have a database of R&D expenditures for 19 OECD member countries dating back to 1973. It is not completely harmonized but one can, for example, see that the contribution of different industries to the growth of R&D in countries is extremely diverse; and it is far better than using individual countries' data.

Internationalization of R&D is certainly an important economic trend and one we would like to capture, but it is not one that lends itself to collecting good data. The OECD is drafting a Globalization Manual that deals with broader issues than R&D, but it has a chapter on science and technology. This manual is nearing completion. We are also collecting data on the Technology Balance of Payments (TBP), which is one component of these R&D imports and exports. To be candid, we are not happy at all with this TBP data because the sources are very different across countries, the classifications are different, and the quality is highly variable. This is something that we might start working on in the coming years if our NESTI delegates consider it important. It is certainly a big problem for our data at the present time.

I want to say a word about the relationship of R&D and national accounts. In my view there is a need for learning on both sides. Not only should those responsible for the national accounts learn from the R&D statisticians but also the R&D statisticians need to increase their exposure to the national accounts officials. The treatment of the data is not always the same. From the point of view of people having to decide a government's budget for R&D the relevant question is how much money was spent last year. For economists interested in the contribution of R&D to economic growth this is not always the best way to treat the data, and some notions such as cost are not necessarily relevant.

Finally, even for R&D policy, it is important to go beyond the R&D data themselves. For example, when you are reporting to your boss in the ministry of research and trying to convince him or her of the best uses of the money allocated, you need some kind of output indicators. That effort is drawing a lot of attention and effort, focused currently on two major sources -- innovation surveys and patents. Further work in this area may affect our R&D statistics.

The next speaker is David Trinkle from the Office of Management and Budget.

DAVID TRINKLE: OMB is the budget office of the White House charged with preparing the executive branch budget, overseeing day-to-day management of agencies, and addressing policy issues that arise. I provide OMB oversight of the National Science Foundation and am responsible for R&D issues across agencies.

OMB is both a data provider and a data consumer. OMB updates the definitions of R&D across agencies; we typically deal in what agencies are authorized to spend but we also report what they obligate each year. We report R&D numbers in a chapter of the budget devoted to R&D across the government. It includes historical tables going back 40 years or so. My side of OMB does not examine the economics of R&D, the performance side. There is an economic shop within OMB and also a couple of White House offices that address that.

First as data provider, OMB is responsible for Circular A-11 setting out all the definitions, for example, of applied development, R&D equipment, and R&D facilities. I will also discuss the federal science and technology budget, a compilation that differs somewhat from research and development spending. We also produce cross-cutting analyses of R&D in broad fields across agencies -- information technology R&D, nanotechnology R&D, and climate change- related R&D. In the budget we show the cost of any R&D tax credit proposals. The last two budgets have addressed the revenue cost of permanently extending the research and experimentation tax credit. And in recent years we have also taken a cut at how various agencies allocate their funds – whether they are subject to congressional direction, for example, or make awards by competitive peer review.

We also are asked for certain other types of information, typically by Congress but also by other sources. This poses a challenge unless we can readily obtain it from other sources. For example, the field of science distribution of spending. We are very familiar with the National Science Foundation Federal Funds Survey data and refer to it quite often, but we not in a position to track that ourselves in real time. From time to time we are called upon to provide specific targeted analyses for which we must rely on the agencies. The analysis of homeland security R&D actually started out as a special call to agencies outside of the budget cycle; now we are collecting that and presenting it as part of the budget. There are other sources for some of these data. The RAND Corporation's RADIUS database is one example.

As a user of data, OMB has a number of different needs and ways of encouraging agencies to use the data. In considering broad agency budget proposals or initiatives across agencies, we encourage proponents to support their requests with data. NSF, for example, identified the need for larger grant sizes to support their graduate scientists and engineers, and they conducted a study that was persuasive. Over the past few years, as a result, we have been pushing in that direction. We have also encouraged agencies to articulate and measure the benefits of their programs and to provide data on performance and impact. In fact, as part of the President's Management Agenda we have asked R&D agencies to justify the relevance, quality, and performance of their programs in making new proposals or continuing existing programs.

Although I am talking about R&D data, I should also refer to the federal science and technology budget which came about in part as a result of a National Research Council recommendation that we needed a better estimate of the investment in new knowledge creation as distinct from the development of new end products or prototypes, for example, military equipment. Federal S&T is closer to what we are talking about when we speak of research investments. We can track this compilation during the appropriations process; we are not able to do that with R&D.

Finally, in preparing the budget's R&D chapter this year we included a number of charts that derived from other data sources. We used NSF data to look at some of the international comparisons. We used data from the *Chronicle of Higher Education* to track the phenomenon of congressional research earmarks. We use these data to help describe the context for the current budget requests, articulate some of the concerns that motivate the requests, and anticipate some of the results we hope to achieve with our budget proposals.

Now as a user of R&D data, we clearly hope to make better informed budget and policy decisions. In this role we also see several of the limitations of the data. One problem is alignment to the structure or level at which decisions are made. Again, for the federal R&D budget, the data are collected at the very last second and nobody makes policy decisions with respect to R&D across the government. The so-called federal R&D budget is really just a compilation of individual program levels so it is something of an illusion to talk about R&D government-wide. There is also a problem with fields of science. Some agencies have large shares of particular fields; for example, the Department of Energy supports much of the physics research in the government. DOD and NASA support much of the engineering R&D in the government; but support of other fields is much less concentrated. So it is hard for us to approach issues in fields by looking across the agencies. Sectors involve a similar issue—universities versus industry.

There is a problem of time lags. We love the NSF data but because they are collected by survey, they are already a couple years out of date by the time we see them. By the time that we are trying to affect policy through our funding decisions, we have to assume what happened over the past year or two plus a year to two before the budget kicks in. So it is hard to know if the concerns we were trying to address are still valid and how we will affect them by the decisions we make this year.

There are obviously problems of consistency across agencies even though the definitions are supposed to apply to the entire government. In each agency those definitions may fit well in some cases and not so well in others. Some agencies find it very easy to draw a distinction between basic and applied research, others seem to have a more distinct line between applied research and development, but that is not uniform. Furthermore, I have concerns about data quality, not so much with respect to NSF data but with regard to data the agencies generate for their own budget justifications.

We see arguments that we accept and are trying to address and others about which we are skeptical. The physical sciences have received a declining share of total R&D, a very common complaint these days, while the life sciences' share has been increasing. We are addressing the physical sciences in the current budget. Yet there are other arguments that we believe are debatable and that we have not accepted as premise of budget policy. It may be that the decline in support of a particular field is relative to a peak that occurred for a reason, and the decline is not a major problem. Nor does the fact that there has been a shift in support mean that it must be shifted back the way it was 5 or 10 years ago or to some baseline the agency argues is appropriate. It is very hard to say at what point everything was just right and we should be moving back toward that point. Different interest groups will all point to different times to which they would like to return, but we cannot move in different directions at once.

To summarize, although it is important to have R&D data, they will not always be available and often they will not be available in time to inform a decision. Yet we want to make the best decisions possible, so the more data we have and the timelier they are the better from our perspective.

The last speaker in the first session is Gregory Tassey of the Planning Office of National Institute of Standards and Technology.

GREGORY TASSEY: I approach this subject from the point of view of an R&D policy analyst. Although they are not followed systematically anywhere, there are four steps entailed in doing R&D policy analysis. The first step is to document and explain the role of technology in the economy. You might think that this is unnecessary, but policy makers are not entirely in agreement about the importance or the specific roles of technology, especially in times of slow economic growth when companies are using R&D and the resulting technology to cut costs and to avoid hiring new workers. Even in the first part of the 1990s and into the mid-1990s, members of Congress were actually questioning whether R&D spending should be increased or in some cases maybe cut. So one of the first uses of R&D data is relating it to the impacts of technology on the economy.

Second, a policy analyst looks for indicators of under-investment. If you do find indications of systematic under-investment then you go to step three, you have to do some kind of cause and effect analysis to relate impacts to something that appears to be wrong with the patterns of R&D investment. And by patterns I mean the adequacy of the amount of R&D and also the composition. If you do your homework here you can then match causes with appropriate policy responses. Getting to step four requires a lot of analysis. It is multidisciplinary and certainly the amount and quality of R&D data are critical. I am going to highlight briefly some of these points, with emphasis on steps two and three.

Any R&D agency, especially one that has industry as its primary client, has to go through this sort of circular flow of economic analysis. One has to find evidence of and demonstrate that there is a pattern of systematic under-investment relative to some optimum, which of course is difficult to define. And then and only then can you provide the economic rationale for the existence of a federal R&D program. Intellectually this is the most challenging use of economics in R&D policy and although economists have made some progress in characterizing and measuring under-investment, we certainly have a long way to go.

If you do make that rationale and then have an opportunity to implement it, more and more OMB and the Congress have told R&D agencies that they have to be much more systematic in the way in which they propose and develop program initiatives. So R&D data have become more important in the strategic planning phase. This is basically sector- or technology-level analysis. If you do your homework then you get a budget. That used to be the end of it, but now more and more you have to come back after the fact and look again at the relationships between R&D investments, including in this case the federal component, and show the kinds and magnitudes of economic impacts that have been realized. And then to close the loop that information feeds back into a restatement and refinement of the original economic rationale for the program's existence.

One of the main hurdles for R&D agencies in accomplishing this is the lack of any consensus model of innovation. Because GDP is the key policy variable in the minds of members of Congress, the R&D policy analyst has to start there and work backward. The problem with this model is that when you work back to the technology investment component, and technology is characterized as a homogeneous entity, it is very difficult to make a clear case for many of the federal roles in supporting technology. Technology is looked upon as a proprietary good in contrast to the science base, which is recognized as a public good. Many believe that if the latter is provided, the private sector will provide the technology. So this causes policy analysts a lot of grief. It is still a model widely adhered to.

For our purposes in R&D agencies, we need to disaggregate that technology box into a set of elements that are not arbitrary but represent distinctly different investment centers, because unless you do it that way you will not get to the right policy prescription. We at NIST have come to use are the following three elements. First is the science base, which is not particularly controversial from a policy point of view because it is considered a public good and therefore government funds virtually all of it. The only issues are how much and how the funds are distributed across fields of science and to whom.

The next element is the technology platform or generic or fundamental technology; it goes by different names. It constitutes the platform from which the many market applications represented by the proprietary technology box evolve. Next is something I call infratechnology, which is a collection of infrastructural technologies present in any high technology industry and that are essential to the actual conduct of R&D and eventually the control of the production. For an agency like NIST, we are obliged to collect our own data to develop and rationalize programs that affect either the generic technology box or the infratechnology box.

As an example of how one would use R&D data at a highly disaggregated level, which is necessary to apply this model, consider biotechnology. First is the science base that has many elements. The infratechnologies, too, are numerous. Generic technologies fall into two categories, product and process. Finally, there are proprietary technologies. The analytical challenge for policy is what types of data you need to support measures that ensure that the industry grows and prospers. The mechanisms that work well in one area of the technology-based economic process may not work particularly well in others.

At the national or macroeconomic level, NSF provides a lot of very useful R&D data, but that is only the point of departure for managing the research portfolio and programs of an R&D agency. At one level the R&D intensity of the U.S. economy has not really changed significantly over a 40-year period, during which many countries around the world have significantly increased their R&D capacity. And as we heard, other countries including the European Union have as their policy goal to greatly increase their R&D intensities. So at a very aggregate level you might say that this indicates there should be some concern about the level of R&D investment in the U.S. economy.

In terms of composition, the federal share of national R&D has declined while the industrial share has increased significantly. This is a trend that is going on in all OECD nations. However, unless you believe that industrial R&D is a perfect substitute for federal R&D, you have to be concerned about this gap as it continues to widen. But, at this level, there is not much more you can say. So we look for indicators at lower levels of aggregation.

And here we break the economy into two basic parts, manufacturing and non-manufacturing, the latter being primarily services. And you can develop indicators that show that manufacturing, which has been largely ignored by policy makers and continues to decline in size, nevertheless accounts for a large share of R&D. This still is not sufficiently disaggregated to allow the type of R&D policy analysis that I have described. So we go down one level further. Now we are approaching the level at which R&D agencies need to analyze trends and begin to plan their programs. Within manufacturing there are so-called high technology industries with much higher R&D intensities than the manufacturing average of about 3 percent. At this level it appears that the U.S. economy has a very skewed pattern of R&D conduct across industries. It turns out spending R&D is also highly skewed geographically. Down to this point all of the data are available through NSF. Beyond it, the question becomes how disaggregated are the R&D data needed to effectively plan and manage an R&D program and how they are obtained at reasonable cost?

All of my comments so far have had to do with the issue of data adequacy. However, the composition issues are at least as important, and here we have even less data. A recent study that shows the importance of looking at compositional changes in R&D over time involved a survey of high technology. It confirmed that most product launches are derived from the current generic technology. In this survey such technology accounted for 86 percent of launches, which in turn drove 62 percent of revenue and 39 percent of profits. The relatively small investment in next generation technologies accounted for only 14 percent of launches but 38 percent of revenue and 61 percent of profits. So the point you make to policymakers is that attention needs to be given to the patterns of investment in next generation technology.

What kind of data do we have? Again going back to the national level, these NSF data show that over the past 10 years development, which is where most industrial funding goes, increased over 70 percent. Basic research increased almost as much, but that is misleading because most of the increase is the result of growth in federal funding for health science. As you probably know from the debate currently going on about doubling the NSF budget, it had not grown much at all in this time period.

In between is the transition phase between basic science and applied research and development. It has grown at a much slower rate. So here again is a rough indication of some national under-investment, but it is necessary to disaggregate not only by technology or industry but also within applied research. The scope of what is defined as applied research is too broad to really match up with the investment trends and the organizational structures of R&D firms.

One of the few examples of an indicator to help answer this question about composition is provided by the Industrial Research Institute (IRI). In their annual survey they ask member firms, accounting for about 40 percent of U.S. industry's R&D, about their planned investment in directed basic research, which is the first phase of technology research, the attempt to apply science. A few years ago the IRI instituted what they call a sea change index for all of their indicators, including this one. I went back 11 years and computed it just to have a longer time series. It is derived basically by subtracting the percent of respondents planning a decrease in directed basic research from those planning an increase greater than five percent. The zero to 5 percent respondents are omitted because that amounts to standing still when you adjust for inflation. Now what you can note here is that every single year this index is negative. This is not an ideal indicator. None of the indicators that I have discussed is ideal, which is why you need a lot of different indicators. You also need to maximize the quality of the data, as well as have the right conceptual framework to define the indicators.

In summary, these are the types of impact measures that an R&D agency like NIST tracks. The broad categories are obviously familiar, but in doing either a strategic planning study or a retrospective economic study, we have to target the selection of metrics to the nature of the program that is being analyzed. At the microeconomic level no one can expect NSF to collect these data because the task is very costly. However, benchmarks of some type would certainly be useful. But we are left with trying to do this pretty much on our own, and that is where I am concluding these remarks. If we are going to do good R&D policy analysis then we have to be very conscious of the amount and the quality of R&D data at these different levels of aggregation.

RECENT DEVELOPMENTS IN NSF'S R&D DATA PORTFOLIO

JOHN JANKOWSKI: My task to describe everything we have been doing to improve our R&D statistics. I can only highlight some of the steps we have taken. In terms of the industrial R&D survey, which is so often the focus of these discussions, we have made some changes to the survey form itself over the last decade. We dropped a few items. For example, we used to collect information on applied research and development by product field and on basic research support by science and engineering field. We no longer collect these statistics from the industrial performers for the very simple reason the item response rate was so low on those survey questions. However, we have expanded the questionnaire to address policy issues. For example, we have moved to collect more information on the foreign R&D activities of industrial performers and on the amount of R&D that they contract out.

The contracting out question serves two purposes. It gives us a better handle on some of the outsourcing activities. It also allows us to look at a difference in what the federal agency reports in terms of their R&D support to industrial performers versus what industry reports as R&D support obtained from the federal government. The Department of Defense reports a lot more R&D going to the industry performers than industry reports receiving from DOD. Maybe this is a case of contracting by one industrial performer to another, and then you lose track of those dollars if the secondary recipient of the funds views them as from another industrial firm rather than from DOD.

Also in the industrial survey we have expanded the number of questions for which reporting is mandatory. This is a unique survey in the sense that most of the questions on the survey are voluntary for respondents. Historically there have been four mandatory questions -- sales, employment, total R&D, and federal R&D. This year we have made the state distribution of the R&D activity a mandatory reporting item. This helps us answer one of the most basic questions about R&D activity, which is where it is done. This year we have received approval to mandate the entire survey form, so for the first time in 50-odd years, all of the questions must be answered by industrial firms.

We have added some questions about collaborative alliances to the survey form and questions about activities involving new technologies -- biotechnology, software development, nanotechnology, and new materials.

We also have been conducting in-house analysis and working with our colleagues at the Bureau of Census to achieve better estimation, for example, of the state-level distribution and composition of the R&D, whether it is basic, applied, or development. These are some of the questions we are most frequently asked about industrial R&D activities.

I want to turn now to university R&D. We serve several masters in collecting and reporting these data. In the university and college survey we have added questions about pass-through funding--when one university received monies and passed it through to another institution. We did that to address the difference between what federal agencies report giving to performers versus what performers report to us having received. It has the added value for giving us a better sense of what kinds of collaboration are occurring in university research.

We also have made changes in the academic survey in terms of non-science and engineering R&D activities. We did that not only because our university respondents were interested in such things but also because other countries include not only the science and engineering R&D but the non-science and engineering R&D in their totals. This will improve the international comparability of our university data but it will also enable us to report breakdowns by institution.

We have totally redesigned our research facility survey of bricks and mortar construction and acquisition, the space where the R&D is done at academic institutions and at biomedical institutions. The survey includes an entire new section on cyber infrastructure, so not only the bricks and mortar but also the information technology infrastructure in support of research. .

On the federal side we have worked together with OMB and agencies on definitions of R&D, in part to reconcile what is collected by NSF and what is collected by OMB. In theory the information that agencies are being asked to report is identical. I agree there is some question about the details because it is easier for some agencies than others to report this type of information. David Trinkle mentioned the federal science and technology budget. We have expanded the NSF survey at least for the Department of Defense to break out the development components. That part that is advanced technology development is probably comparable to what a lot of other agencies do for development versus major weapons systems development.

We occasionally have forays into nonprofit R&D surveys and state government R&D surveys.

A few years ago the Division of Science Resources Statistics did a customer satisfaction survey looking at all of our products—R&D expenditures, R&D infrastructure, science and engineering workforce, science and engineering education, and pubic attitudes toward science data. Of approximately 112 responses, about 60 rated expenditure data most highly. Next was R&D infrastructure with 17 responses. We have now begun to look in more detail at uses and users of our statistics. We contracted with WESTAT to conduct electronic focus groups initially involving government officials and then a random sample of SRS data users.

A few themes emerged. First, the integrity of historical trend data is important. The problem there is that if we continue to use surveys designed for circumstances 40 years ago they may not be as relevant as they ought to be even though we have preserved a good time series. Although generally people were happy with the accuracy and the comprehensiveness of statistics, many complained about their timeliness. I definitely agree that that is a problem.

On other issues users would like to see more industry detail, more information on multidisciplinary research activities and on emerging research fields, and more information on sub-state activity. Our effort to collect data on the state distribution is partly responsive to that demand.

We have asked the Bureau of the Census as part of the general survey collection activity for industrial R&D to do some special studies and investigations for us by posing a few additional questions to businesses. For example, we have asked how good are the definitions that we provide for capturing service sector R&D. The response to that question was "not very bad but not necessarily good." We have since funded a more detailed investigation that I will describe. In the meantime, we changed some of our definitions, for example to achieve better coverage of software development and clinical trials.

What may be most relevant to today's discussion is that we have asked businesses about providing information on R&D below the company level even though we discontinued collecting product field data for applied research and development. Of the 45 companies interviewed, 43 could report estimates of R&D spending at a sub company level, but there was a great deal of variability in the level for which reporting could be done. At the division level 6 firms, at the subsidiary level 9 firms, for line of business 6 firms, for product line 8 firms, by cost center 5 firms, and 11 others at some other disaggregated level. So our conclusion is that at this time we should not recommend that companies be asked to provide estimates of R&D spending at a sub company level. But we will no doubt revisit that issue in the future; a negative answer does not mean that we won't pursue it.

With regard to the service sector we have funded with NIST a project to look at taxonomies and planning issues for services sector R&D. Mike Gallaher here from RTI is conducting the study for us. In the early 1980s services R&D accounted for about five percent of what was reported on the industry survey. It is now up to about 36 percent of the total. We have several questions about what we are actually capturing, so we asked RTI to look broadly at the sector but also incorporate case studies of the telecommunications industry, software development, financial services, and research development and testing services.

Another study is looking at fields of science, especially levels of aggregation in our surveys, how to track multidisciplinary research, and how you array the data. A paper that I wrote with my colleague, Francisco Morris, and presented at Statistics Canada addresses data on alliances, research joint ventures, and technology transfer activities. In these cases we are engaged not in collecting a lot of statistics but in serving as a clearinghouse for the government. We work with other agencies and a lot of private organizations to tabulate things and to bring them to the public eye.

Finally, a topic that will be addressed in a later session is cross-national R&D. The National Science Foundation, Census Bureau, and the Bureau of Economic Analyses are about to sign a memorandum of understanding to link at the micro data level the statistics from the industry R&D survey with data that BEA collects on foreign direct investments in the United States, and the U.S. direct R&D investments abroad because two BEA surveys collect R&D statistics. This will be described in more detail later on today.

COMPOSITION OF PUBLIC SECTOR RESEARCH

BILL BARRON: The subject of our panel is composition of research and development expenditures and activities, a difficult and complex matter as many of our panelists have already suggested. We will focus first on the public sector, federal and university research, beginning with a presentation by Michael Saltzman from the Department of Energy.

MICHAEL SALTZMAN: I am going to address two general topics, one is how we allocate funds among basic research, applied research and development, and the second is how we allocate funds among the fields of science. Let me tell you a little bit about DOE. DOE is a $23 billion dollar agency in 2004, and of that R&D accounts for about eight and a half billion dollars. We are organized into four general functions -- defense, science, environment, and energy. Energy is only 11 percent of the total budget but yet we are still called the Department of Energy. Out of the $8.6 billion, most of it is in the science and energy areas, defense and environment are less percentage wise, they are more product oriented.

Just a few statistics. The DOE R&D budget is the fourth largest in the federal government after DOD, HHS, and NASA -- third largest in basic research, fourth largest in applied research, third largest in development. We have the second largest budget for construction and capital equipment, our accelerators and light energy sources primarily. We are the third in spending for information technology, the second in spending for nanoscience, and we have a substantial program of climate change research.

Before I talk about how we allocate the basic and applied research and development, let me talk to you briefly a bout when we do it. In November of every year we get ready to report to OMB on our R&D funding. In November of 2002, not so many months ago, we reported on 2002, 2003, and 2004. But the only reliable data at that point were 2002 because the year was just completed. For 2003 we didn't yet have an appropriation and for 2004 we had only a budget estimate. Unfortunately, in the Department of Energy, we have no data systems that provide us that information; we have to go out and collect it from our program secretarial officers. A lot of the data are very last minute. David referred to the fact that some of the decisions are not made until the very last minute in January and it could very well be that what is reported in the President's budget does not actually reflect what is really in the President's budget because the OMB data system is locked up in early January but changes may still be made for another week or two after that.

I also wanted to mention the SBIR system: SBIR was authorized in 1983, and it has been reauthorized several times since. Currently 2.5 percent of extramural R&D is to be directed to small businesses. Extramural R&D at the Department of Energy means anything that is not done by a DOE federal employee at a government-owned government-operated facility. Most of our defense related research is excluded from SBIR. I am telling you this because we use a cascade to get to the SBIR number and this relates to how we check some of our R&D numbers. We start with a total program funding and we determine what part of that is R&D. Then we look at the R&D and we determine what part of that is extramural. Again, it's a checkmark and we have trends to make sure we're doing that right. And then based on that we can take the SBIR tax.

As we do the calculations we do find problems. Sometimes they are recurring problems from year to year, and sometimes they are new problems. And in some programs there are incentives to underestimate R&D because then you reduce your SBIR tax. We take care to try to ensure that that does not happen.

Now a few observations about how we allocate R&D among basic research, applied research, and development. Let me just simplify how we look at it. Basic research I would say is knowledge driven research; applied research is need driven, and development is product driven. I will give you two extremes. The first is the DOE Office of Science, $3.3 billion of basic research. It is high risk, long-term research of the kind that you might expect a university to be doing. But it requires major facilities such as accelerators, light sources, and reactors that are currently running a billion dollars and more to build. The International Fusion Project is going to be a $5-plus billion project. It is clearly research without a specific application in mind and therefore unquestionably basic.

On the other extreme at DOE is the Office of Naval Reactors, which develops naval nuclear propulsion systems. This is very product oriented and if you were to look at the OMB definition of development -- production of useful materials, devices and systems –that is where it would fall.

Then we have other programs that fall someplace in between and their characterization is a bit more judgmental. Our weapons program, for example, has evolved from the design and testing of weapons – a development program --into a more applied research program with stockpile assurance and modeling and simulations. The nuclear energy program ranges from basic through applied research through development depending on the stage of the R&D project.

In environmental management most of the R&D is development, and that is true of the fossil energy program. A few years ago we undertook something called portfolio analyses, which gave us a tool of analysis at a project level and a means of classifying projects as basic or applied research or development and determining what field of science they were in. In short, with regard to the allocation among basic, applied, and development, I think we are generally doing a good job, although some offices are doing better than others. Mainly that is because research projects are pretty clearly in one category or the other.

The second topic is the allocation by field of science. In 2001, DOE had a total of $4.6 billion in basic and applied research and another $3 billion in development and capital equipment and facilities. Of the research total 68 percent was in just five of NSF's twenty-plus fields of science -- biology, chemistry, physics, computer science, and metallurgy and materials. Over the weekend, I reviewed some of the allocations by field, but instead of looking at the budget and deciding where the money went, I looked at the NSF data and asked myself if I could figure out which programs the figures represented. I was quite successful in doing that, which suggested to me that the program offices have generally been doing the classification correctly.

If you compare the 2002 NSF field classification and the 1982 classification you find that there is virtually no difference except for a couple of words. Does it make sense that there has been so little change over 20 years during which, for example, there has been a revolution in biology with the human genome project and the emergence of computational biology? Furthermore, there have teen presidential initiatives in nanoscale science, information technology, and climate change as well as the growth of interdisciplinary research. A second issue is that development is not captured in fields of science which may be a problem when we do the allocations between applied research and development, which are not exact.

Here are my conclusions regarding fields of science. First, I strongly support consistency in the classification. I am not advocating any changes; we like the fact that it stays the same from year to year. But it does become more problematic, there is some sacrifice of accuracy as computer modeling and simulations become more equal partners with experimental and theoretical research. We should be able to take into account, for example, that in the DOE weapons programs modeling and simulation are replacing physics.

We have also had to address what happens when computer science programs directly support other research. We have advanced computing programs that are supporting nanoscience; climate change; and fusion. When we report in the OMB crosscuts, we show the computer science as part of nanoscience or climate change. But when we respond to the NSF survey, we report it as computer science. It may not be a good thing that we are not consistent. How should we address the fact that the NSF survey does not account for presidential initiatives? I do not thing that we should modify the fields of science. I would much prefer to use crosscut-like entries for subfields within presidential initiatives even if they were not consistent from year to year nor completely accurate. The data would nevertheless be captured.

And what about the fields that were not even heard of 20 years ago, like the human genome? We now pick it up in biology, but that is not giving it its due. Much as I support the consistency of the federal funds survey it seems to me we ought to be calling in experts to make some cosmetic changes and bringing it up to date.

The next presenter is Charlotte Kuh from the National Research Council.

CHARLOTTE KUH: I was asked to discuss this topic because I am directing a project that is trying to classify research fields for the assessment of doctoral programs, sometimes known as the National Research Council (NRC) rankings. This is somewhat different from your problem, but in fact there are similarities. Research and development is typically carried out by scientists and engineers, many of them with doctorates. And the R&D dollars that flow to the academic sector may have a significant effect on the training of doctorates and their quality. I'm not sure, however, that the doctoral education community wants the same thing from a field classification for doctoral programs that the R&D community does. Finally, the R&D statistics do tell us about resources to the academic sectors, but they don't tell us that much about bodies.

This is what the academic community is looking for from a taxonomy. They are trying to classify doctoral programs to obtain reputational rankings of similar programs and to associate a wide range of descriptive data about students, faculty, and research resources with each program. This association will inform us about whether those programs are deemed of high quality. They train a large number of international students, and whether these programs also receive a large amount of federal R&D funding. We are not seeking fine detail about the kinds of research being done in these programs; we don't know who is training the telecommunications researcher of the future, although we are interested in computer engineering programs and electrical engineering programs. Success on our classification problem would mean that we have grouped similar programs and put dissimilar programs elsewhere.

Furthermore, we are taking a student-oriented point of view. We want to tell potential doctoral students what is available to them in terms of emerging fields and interdisciplinary centers.

We also have some secondary objectives. If possible, we want to maintain some continuity with the past and to use categories that will be meaningful 10 years from now if we are lucky. We don't want categories to be too detailed because we want to compare programs. On the other hand, we don't want large super categories where a rater would be unfamiliar with the random selection of programs from that category.

We are looking for data from over 300 universities and probably 3,000 programs. Each university has its own nomenclature, but we need to give names to our categories so that a university can tell whether it has a doctoral program and in what category that program should go. We do that not only through these broad fields but also through subfields. If we were to try to use the NSF classification that Michael referred to or the NSF personnel taxonomy for that matter it would not work simply because those are 1980 categories and we are now in 2003 and things have changed.

Then there is the problem of interdisciplinary research. Fortunately the names of doctoral programs don't change as rapidly as the research that is carried out within them. Further, we can ask faculty members if they are involved in doctoral education in programs outside their primary program, so that will give us some notion of cross-disciplinary research while finding out both who hired them and who pays their salary.

Our work is still in progress, but let me describe three instances in which we have tried to improve the taxonomy of doctoral programs compared to the taxonomy we used in the last evaluation. First, the biological sciences. We really messed up in this area the last time we did our study, and the community let us know it. These fields are changing very rapidly and the nature of research is changing. Molecular biologists still exist and biochemistry still has departments but now there are also biophysics and structural biology. We have simply added those sub-categories. Similarly, in cell and developmental biology we have three categories including microbiology. Genetics has become molecular and general genetics. Neurosciences now include neurobiology. Pharmacology now is pharmacology and toxicology. We have been trying these areas out on pilot sites who are willing to put up with our data requests; they say well, this is better, but it is not perfect because the biological sciences are highly dynamic.

We also tried to anticipate some field consolidation. We thought that there was an emerging field called geosciences because systems modeling is common to the earth sciences, oceanography, and atmospheric science. I must have heard from every admiral in the world that no, oceanography is a separate science. So we thought about it and now have two categories. One is earth science and the other oceanographic and atmospheric science, partly because the pace of change is faster.

Finally, how do you find emerging fields? Here are three of them -- nanoscience, computational biology, and some parts of the humanities and social sciences such as gender studies and global area studies. But these are really too small for what we are looking for.

Let me tell you what kinds of problems we have encountered with emerging fields. Programs and individuals will try to fit themselves in somewhere if you give them a list. For example, there is a field called evolutionary biology that we initially eliminated because it isn't in the NSF taxonomy. We immediately started hearing from that community and tried to figure out how many PhDs came from that area. We said maybe you exist but you're too small. It turned out that around 80 PhDs have written in the field on the NSF form in the "other" space, but in fact departments of evolutionary biology had graduated over 1,000 students that they thought were in their field. Those 1,000 when confronted with a form that didn't list their field assigned themselves elsewhere. So it is probably difficult to detect emerging fields from what people answer to the category "other." We are going at it by working with the professional societies and trying to find out where these fields are.

Second, fields don't all organize themselves the same way. They are idiosyncratic in how much lumping is acceptable. Chemistry and physics are quite willing to stay monolithic despite encompassing diverse subfields, while the biological science fields will set up a new field with a new name on short notice. Genomics morphs into proteomics, which may be regrouping into structural biology. To work, a taxonomy needs to permit field specific variation and the fineness of disaggregation, because that is just the way the fields are. Continuity is important, but sometimes we need to break it. But regardless of whether we change it, we should review the appropriateness of a taxonomy with some regularity.

Finally, interdisciplinarity. I am trying to figure this one out. Everyone says that interdisciplinarity is pervasive in contemporary research and I believe that claim. But does that mean that our classifications are meaningless? Because if you call something interdisciplinary, what do you call it? We are trying to deal with it by saying people can allocate their time across different doctoral programs, and the total has to add up to 100 percent, and then we can find out people who are supervising dissertations in more than one field. I don't know if that will work for research and development expenditure data collection, but you might think about some means of letting people talk about all of the different areas in which they are doing their research.

COMPOSITION OF INDUSTRIAL R&D

MR. BARRON: We will move now into the business sector to discuss industrial R&D, and our first presenter is Charles Duke from the Xerox Corporation.

CHARLES DUKE: Although I am a practicing scientist, my day job for the last three or four years has been vice president in charge of the research, development and engineering at the Xerox Corporation where I have been active in the management for 20 years. I think I can contribute best to the discussion by giving you a quick overview of the realities of how we do budgeting and planning at Xerox

At Xerox Corporation we manage essentially four things. On the one hand we manage budgets and investments and expenditures, and we do that by a budget process. We manage outcomes for the business by a different process that we call it managing value chains. From my point of view that is the most important determinant of the success of a research project. It is largely irrelevant how good the research is, if it does not have a complete value chain it will have no value. We manage that by something called the "time-to-market process," which is asynchronous with the budget cycle.

We manage individual research projects by something called the "PEP or performance excellence process," whereby at each year each manager determines what his objectives are and these objectives are mapped into programs, and those programs are mapped into quarterly deliverables, and those quarterly deliverables are measured every quarter and everybody's pay depends upon whether you deliver what you said you were going to deliver. You can get your pay doubled or halved easily by not performing according to what is asked of you so we have a very heavy incentive compensation system that is tied totally to whether you deliver what you said you were going to deliver that year.

Finally we have activities. Those activities are determined in the PEP process because the first thing that is decided is what the deliverables are. We allocate budget against those deliverables, and we do spend against the deliverables. Every quarter we reshuffle those expenditures because every quarter roughly half of the activities are working fairly well, a quarter of them are not working as well, and another quarter of them are in the tank. Every quarter we are reshuffling to make sure that we get through the year with a certain number of things that have been completely done and not everything has tanked.

So we measure these four things. We measure how much money we spend, we measure what are the business outcomes of that, we measure what are the outputs that are required to generate those business outcomes, and we spend against the activities that are required to generate those outputs..

Now with that in mind, let me make a few comments about what decisions we make and what kinds of data we use to make those decisions. At the corporate strategic level we do budgeting of investments against outcomes. Those outcomes are commitments on the part of the business manager. As I indicated, their pay depends on it. We separate ourselves into lines of business and in each line of business we make commitments to market share and market share growth each year. And the outcome of that basically is a budget and a high level description of what we want to accomplish for the year, the high level PEP.

At the operating level we take those budget numbers and apply them against specific outputs. And as I indicated, there is a scheme for collecting those outputs into the ultimate deliverables, which are products or services. Xerox perceives products and services as essentially parallel, slightly different but parallel. These products and services are split up into value chains, and each segment of the value chain has deliverables for that year and we spend in each segment of the value chain so that that segment of the value chain produces its deliverables for that year. That is the way we spend our money and what our operating line managers are responsible for.

The RD&E expenditures are an outcome of this process. At the corporate level they are highly aggregated numbers. Over the past 20 years we have spent between less than 1 percent and more than 6 percent of our financial operating revenue on RD&E. That is more or less a target that is set at the corporate level. Now how that money actually gets spent is determined by the needs of each year and there can be wide fluctuations in how that money is spent from one year to the next.

How does this description relate to the classification of R&D? In this scheme of things you see we invest money against outputs, not against basic research, applied research, and development. But can we give you statistics? Absolutely, but you would not have any idea of what we were doing. You could not discern the processes that are going on underneath them that drive how we make the investments and what we expect to get out of them.

We were asked was about the global distribution of funds. Xerox is a global company, and each business unit is global. The unit that I represent, the Xerox Innovation Group, is global with laboratories in Grenoble and Toronto and a joint venture with Fuji Xerox in which we spend about one-third of our research funds. And except for Fuji Xerox, because of its status as a subsidiary, none of those numbers are disaggregated unless they are required for federal reporting purposes. So we are investing against our deliverables, and we don't ask was the work done here or was it done there. We have a set of goals for the year, we figure out what we have to do to accomplish those goals done, we pick the people who are supposed to get that work done, and that is all paid for out of one pot of money.

BILL BARRON: Our next speaker is Bill Long of Business Performance Research Associates, Inc.

BILL LONG: I am going to talk about data on corporate lines of business and new technologies; these are slightly different, although not unrelated, issues. First a brief summary of bases on which research and development data are collected. On the business side, the basic unit, as we have heard, is the project. Someone submits a written proposal and it goes through various stages of approval. Second, there is an R&D plan, sometimes embedded in a commercialization plan and sometimes developed by different people. Expenditure data are reported twice, first to the Securities and Exchange Commission (and made available by Compustat).and second to the NSF in the RD1 survey, starting in 1972. Formerly the data were also collected from members of the Industrial Research Institute by CIMS. Finally, for a time, data were collected from firms on a line of business basis by the Federal Trade Commission (FTC). I was responsible for the latter program for its brief existence between 1974 and 1977. All of these are micro data sets that have been used in a variety of ways and by a variety of researchers over the years to investigate one or more aspects of R&D, whether productivity, return on investment, or public versus private returns. Some of these data sets are subject to restricted access.

In the FTC effort to collect R&D along with other business information, the instructions were to classify activities into business units. The companies were selected for their activities in R&D. They were selected because they were leading R&D performers, about 500 companies in all. R&D expenditure was one of a number of data items collected.

Let us compare the FTC and IRI efforts. The IRI survey asked respondents to base their reporting on their 10K reports, which are limited to 10 business segments. So that the average number of lines of business in 1993, the first year for which I have data for the IRI data set, was 2.3. There was no limit on the number of lines of business that could be reported to the FTC, so the average there is 9.5.

A note about line of business results versus results when companies are assigned to a single industry category. It turns out to make a lot of difference. What I did was to compare the 10 top R&D-performing industries from the FTC publication for 1975, with the NSF report on the same industries in the same year. The number of industries in manufacturing for the FTC approach was 260 while the number of industries in the NSF approach is about 30. So a way to envision this is that the enterprise data, which is the basic reporting and classification unit for the NSF survey, is a weighted average of the lines of business for that company, whether two lines of business or 22 lines of business. Whatever it is, the ratio of R&D to sales for the whole company is, by definition, a weighted average of the R&D to sales ratios for the lines of business of that company, with the weights determined by the sales. So one of the things we can notice is that in every instance the R&D to sales ratio for those 10 finely defined industries is lower in the NSF data then in the FTC data. That's virtually mandated by algebra. But not only are they all lower, their rankings are not quite the same.

Next, I created a diagram of global companies that report to the SEC. Break these down into two pieces-the U.S. part and the foreign part. The U.S. part is what RD-1 collects data for. Break that down into U.S. operations, foreign sub operations, and foreign contract operations. Then further break down U.S. operations into internal R&D activities and contract R&D activities. And finally note that the FTC lines of business are a disaggregation of this piece of the RD-1, the IRI-CIMS business segments are a disaggregation of the global company. So these business segments are global entities, these are domestic only. And that makes a difference if you try to do what Bronwyn Hall and I did in a report for NSF, which was to look at the two micro data sets and try to make sense out of the data that are reported in one place versus another.

Now I want to comment on the study referred to by John Jankowski done by the special services branch of the Census Bureau. As he noted, it does not recommend that companies be asked to provide estimates at the sub-company level because the terminology that is used to describe those sub-units varies a lot, because companies say they are not able to describe units in industry terms, and finally because details on basic research, applied research, and development are not available for such-units. No additional study to address this challenge was suggested although additional study was suggested for the other major subject of that study, service sector R&D.

The IRI-CIMS survey was voluntary and the FTC survey was mandatory. Nevertheless, in both instances companies did assign their sub-company units to industries. In any case, the Census Bureau itself assigns industry codes to the U.S. domestic enterprises. It doesn't ask companies to do that. There are several people in this room who could assign such entities to industries. That is not a difficult exercise if you have enough information on what the unit does.

Data on all three components of R&D, basic, applied and development, were reported separately in both of those actual historical programs. The FTC asked the company to report basic research as a single item for the company and to report applied and development for each line of business. In the IRI exercise, all three items were collected for each of the corporate sub-units reported. No apparent use was made of available Census data. On other aspects of the companies in question, there are data that can be integrated before you ask those questions about their R&D activities. The same is true for external data such as patents in Compustat data.

With respect to capturing data for new technologies, some progress has been made, for example with respect to biotechnology and software development in the most recent RD-1 survey. But these are not treated as industries so there are no sales data or data on scientists and engineers. Do we have a methodology for capturing the next new technology that comes along, say in 2006? As long as the Census NSF approach is to assign each company to only a single industry, there is going to be a huge problem in tracking new technologies.

Consider the dotcom companies. Was it a new technology in the 1990s? What about the R&D component? What was it? What effect did it have? What about cell phones? There is no question that cell phones are a new technology. Do we know anything about it as a result of NSF activities in collaboration with the Census Bureau? No. What about biometrics identification technologies involving newly developed sensors and software algorithms for matching?

My recommendations are as follows. Include fewer items in the survey and make all of them mandatory. I was delighted to hear that this will be the case in 2002. I urge that that be permanent. Make definitions clearer. Handling multinational companies is problematic because a company will report data differently on the RD-1, merely because of an internal legal reorganization of the company structure.

Revisit that Census Bureau conclusion that line of business reporting should not be done. And finally on new technologies–develop a system for dynamically tracking them. It's not hard to do that. Just read the *Wall Street Journal* or the *New York Times* or even occasionally the *Washington Post*. And there are others who have been doing this tracking, so this is not an impossible exercise.

BILL BARRON: Our next presenter is Ron Jarmin from the Bureau of the Census.

RON JARMIN: Our data wish list has already been presented by a few people this morning. We want inputs and outputs on R&D and we want them by industry; we want them by geography; we want to know about partnerships; we want to know about outsourcing; we want to know about international collaborations. That is a pretty tall order for any single survey instrument.

If I were considering that wish list for just a generic activity such as making cars, the place that I would go look for that information would be the economic census. That's the most detailed information on inputs and outputs that we have in this country. And it varies of course by sector. Some sectors such as manufacturing do a very good job. Some people think we should be better, but we collect very detailed inputs and outputs every 5 years in the economic census.

Now under NAICS we collect less detailed inputs and outputs on establishments called R&D labs, so there is actually another source of information that we haven't really talked to here today. Under NAICS we now classify R&D labs and we send them a form tailored for R&D labs and it asks them certain questions. We did that in 1997 and in 2002 we added some product information.

I participated in a trilateral committee on the North American Product Classification System for identifying products of R&D labs, which is a difficult task in itself. As part of that exercise, simply for my own gratification, I took the RD-1 data and I ranked it by the top 100 R&D performing enterprises in the United States. I wanted to know how many of them had what we would identify in the economic census as an R&D lab. It was less than 20 percent, my recollection is somewhere between 11 and 17 percent. The vast majority of the top R&D-performing companies in the United States do not have an establishment that we would recognize in the economic census. In other words R&D is being done in other types of establishments. That speaks to the difficulty of determining where is R&D done, how can we locate it. It is usually not done in a specialized facility devoted to R&D. So we have the RD-1 and the economic census. Both collect some information on inputs and outputs, but the two are not necessarily coordinated.

How did we really want to collect information on R&D? I am going to be a little bit provocative here and talk about a different way of collecting some of this information. I haven't thought about this a great deal but it is one way of thinking about how we might go about collecting the information we want.

In principle we would like to have project level data. Now if we were to try to collect project data from a survey sample things such as the RD-1 the Office of Management and Budget would run us right out of town. That is far too excessive a burden on respondent firms.

In fact it may not be too great a burden. We actually collect data on shipments of goods in the United States in something called the Commodity Flow Survey so that we can track transportation and related trends, and for that we actually contact business establishments and ask them to report shipments that they have made. Similarly, we could contact companies and ask them to randomly select "X" number of R&D projects and then provide some sort of detailed description of those projects. It sounds like companies keep some information at this level, so we might want to go down that road. I think that would help us get at both the detailed inputs and the sort of business outcomes expected from these inputs. You are never going to get this from a census approach nor by sending the RD-1 form to what I believe are primarily accountants at the company headquarters, nor from high level R&D managers. You want to get down to a more detailed level. That is one approach to think about, to survey particular projects within companies. I think we could draw the frame for that from the RD-1 as we know who the big R&D performers are.

Another development at the Census Bureau that I think may help us quite a bit in the future is the advent of employer/employee matched data sets. The longitudinal employer household dynamics program at the Census Bureau is linking together data on establishments and their workers, and I think that offers a lot of useful information about the R&D process and how it differs from one establishment to another. One can track actual individuals from one place to another and learn a lot about other inputs and how those add value to the product.

One more thing I think that we have learned is that the value of any particular data collection exercise or survey is greatly enhanced by the ability to link it to other sources. So for example, the RD-1 although useful in and of itself is even more valuable if linked to the census of manufacturers. We expand the number of issues that we can examine especially in terms of productivity and other business outcomes that you expect R&D to be influencing.

BILL BARRON: The next presenter is Michael Gallaher from Research Triangle Institute (RTI).

MICHAEL GALLAHER: I want to talk about a project at RTI, sponsored by NSF and NIST that is looking into service sector R&D. Our collaborator is Albert Link, University of North Carolina, Greensboro. The motivation of the project is that most of what we know about R&D, how we model it and collect data on it, is based on a manufacturing paradigm. Now that we are seeing that over a third of industrial R&D is being done in firms classified in the service sector, the question is what are firms actually doing, what type of R&D? We hear comments that developing countries are quickly becoming service orientated economies, so this R&D is going to be more important not only here but also around the world. The objectives of the study are to assess the existing taxonomies in service sector R&D and recommend improvements in the definitions and classifications and even methods of data collection.

To focus the study we have targeted on four specific areas in the services sector. We are looking at telecommunications, financial services, computer system design, and research and development and testing, the equivalent of outsourced R&D. In the early phases of the project we are taking a look at the composition of R&D to learn what these companies are basically doing. Then we narrow the focus a little more to look at how the R&D is targeted at the provision of services.

To investigate the composition, we began with publicly available sources, such as Compustat. We took a look at what companies are doing R&D, and then we went to their websites and their 10K reports, and we looked at the trade organizations they participate in, and asked what really is the R&D that's being done by these organizations. And we found some interesting things. For example, in the financial service sector, one of the first things that jumps out is that a lot of the companies classified as financial services are really patent holding companies that are umbrella organizations that manage portfolios of patents and small R&D companies. Similarly, in computer systems and design you have hardware research going on, you have software development. And then you have the growing sector of systems integration. And then with the RD&T sector, of course, you have the trends in outsourcing to these companies, and here you have a lot of small biotech companies that are in fact supporting the pharmaceutical industry. They are here because their R&D is classified as a service.

So focusing in a little more on the R&D that is targeted at the provision of services, we see some common themes across the industries, that important issues are reliability and access and security. And information security is important in almost all of these service sectors. The telecommunications industry addresses these issues by doing R&D both in physical devices and in systems integration.

In computer system design systems integration raises many of these issues, a key one being interoperability. More and more as we move to supply chain integration, linking financial systems with inventory and manufacturing, companies need to have their data systems talk to one another. Focusing on this process might yield a model of innovation in the service sector. This is one of the things that a lot of people have attempted to do, but it still needs work. And so I'm sure something like this, not to say that this is the end-all model of how information flows throughout systems integration processes, but just to show this is what we'd like to have. This sort of gives you a distinction between what you have in systems integration are the application of knowledge, tools, processes, to the developed, at least to the provision of a service which is the integration of these systems. So what you need to ask is where R&D-- will and in the broader sense innovation-- enter this system, and how are we going to measure it.

Two issues come to life. The first being that when we went out and began talking with companies that specialize in systems integration, we got a variety of answers of what type of R&D do they do, and whether they do R&D at all. Some companies view systems integration as the development of customized, one-of-a-kind products, and so from that perspective almost everything they do in the developmental stage they will call that R&D. Now in contrast, we talked to other companies who say well, we don't do R&D. We have this package of tools and knowledge, that we use these tools to repeatedly provide a service to companies. So from that perspective we're just doing the same thing over and over again. Now those are the two extremes, and I think a lot of companies will feel that they fall in between these. So you have the big companies, IBM and Hewlett Packard, for which a large share of total revenue is R&D. But then you have other companies conducting large amounts of system integration that report they do absolutely no R&D. So therein lies part of the problem.

Now from our informal interviews with some of these companies, we asked them well, if you are doing R&D, what type of R&D are you doing, what do you consider R&D? And they tell us that building new systems requiring a significant amount of software development, they consider R&D. They also count maintenance of code as R&D even though it is not really encompassed by the survey. But respondents are referring to the expandability of these systems and in effect they are morphing into new products. I think that is what they're thinking about. You also have something you might term applied R&D, developing better processes and establishing generic protocols that can be used in a wide variety of applications across different businesses.

So in the next phase of our study, we are developing some models of innovation that can be used to help us build a taxonomy that will resonate with industry. We want to help industry identify what is R&D and what is not, and then go out to them with a small sample and see what they think about that taxonomy, and get back questions and modifications. The eventual goal is to make some recommendations for the instrument in two categories. The first category is simply better examples and definitions up front to help industrial respondents think through the issues. A prime example is customization. How should they think about customization in terms of R&D? The second category of recommendations involves modifications to the instrument itself. Here what you might have are some questions on technology areas, such as biotechnology, nanotechnology, and areas of software development such as systems integration. This will be important guidance for firms in the manufacturing as well as in the service sector because both are undertaking these types of activities.

BILL BARRON: Our last speaker on the topic of private sector R&D, primarily in services, is John Alic, consultant and formerly an analyst with the Congressional Office of Technology Assessment.

JOHN ALIC: The theme of my remarks is "Let's not be fooled again." We were fooled in our view of the service sector for a long time, but we have a chance to get it right and we should take advantage of that opportunity.

I want to make three major points this morning. The first is that technology has always been important in the services back in history to the era, for example, when railroads designed and built their own rolling stock. Second, today we do not understand, except in the very crudest way, how technology functions in the services. But a little reflection should persuade us that it's much more than just computers and IT system integration, those are certainly very important as is the synergy between goods and services, as in the railway example. However, as Mike Gallagher indicated, innovation in the services does work fundamentally differently than in the manufacturing sector. It is possible to figure that out and design indicators based on that, and we should begin to do that.

The third point is that to accomplish that task requires really digging deeply into the way service outputs are produced in the economy and working from that back upstream into the technology. I think when we do that we will find that conventional notions of R&D are not especially useful. That doesn't mean that they are not of great historical importance, particularly for seeing what is happening to the goods produced in sectors of the economy, but technology and technical activity in the services are really quite different. Services have always been of great importance in our economy. Ever since we began collecting statistics the service sector has been bigger than manufacturing, but plainly that dominance is rising.

There have been insightful voices telling us for a long time that technology is important in the services. Victor Fuchs is perhaps best known for his work on health care, but he wrote an excellent book in 1968 called *The Service Economy*. Nathan Rosenberg wrote about the air transport industry which is one of the stellar performers in terms of productivity growth going back to the 1930s. Both pointed out the way in which users have pushed along the process of innovation by feeding information back to producers. These are ways in which technical activity takes place. In many service sectors it has gone on for a long time and is still going on. There are many, many examples in other service producing industries that we could look at.

I don't know what service R&D was in the 1980s anymore than we know what it is today because there is a huge error band in the nonmanufacturing R&D numbers. But if we were to back extrapolate the more recent figures we would see that it was substantially larger than the statistics indicated.

What does this mean? It means that R&D is a cost accounting convention. It means that the convention, whatever it is, needs to be applied more consistently across the economy. We should accept that R&D is one slice of an amorphous, fuzzy set of technical activities that we don't understand very well and define it operationally. That operational definition is perfectly respectable scientifically. What is R&D? It is what the surveys say. What is IQ? It is what an IQ test says it is. That's what the number means, nothing more, nothing less.

We know very little about what goes on in terms of technical activity in the services. We do know that there are a lot of scientists and engineers employed in the services, in fact the majority of them for some years now. This should have told us that we are missing something. We would be far, far ahead if we had disaggregated data by occupation in the science and engineering occupations in a way that we could correlate with industry of employment. We don't know very much about that. If we did we would have better understanding today. Looking ahead we should try and make sure that 20 years from now we have what we need to know about what's going on.

Some of the available results are implausible. John Jankowski indicated that the figure one of one- third of nonmanufacturing R&D for wholesale and retail trade is some sort of glitch. I will simply note that the figure for health care services is implausibly low. In case I'm wrong and the trade figure is not a glitch, you still have to ask yourself is it really plausible that finance and insurance account for less than a sixth as much R&D as the trade sector. I don't think so.

The fundamental differences between technical activity in the services and technical activity in manufacturing reflect not only that people do different things in the services, but also fundamental in the way that technology is taken up. In manufacturing the product, which is the technological artifact, is fully specified. The major specifiers are professionals of one sort or another–manufacturers, engineers, scientists. They pin down the attributes to that product. They tell the process people what it is. There are process sheets that attempt to enforce strict conformance between the intended specifications, the technical specifications, and the output that is produced. None of that with a few minor exceptions such as fast foods really holds in the services even for most banking products. In the services, the output is produced at the time of production. It does not have a fixed predetermined identity. The technology is more often a facilitator than the result itself. And when we begin to understand in more detail how this works, and feed it back into the technology and knowledge creation activities in our economy, then we will have a better conceptual framework for R&D in the services.

NICK VONORTAS: I do know from professional experience that the Europeans have done a great deal of work on defining services R&D. There are research units dedicated to the issue. How much are they integrated into the activities of NSF?

RON JARMIN: I would characterize the European efforts in terms of services R&D in the same fashion. They are influential in the OECD's development of the *Frascati Manual*, the guidelines countries aspire to for collecting R&D statistics. France has been a lead country there. But there have not been enough successes that could be generalized across countries in terms of the redesign of R&D surveys either their definitions or instructions.

R&D COLLABORATIONS

DONALD SIEGEL: I am going to begin the session on R&D collaborations with some brief background information for those of you who may not be familiar with the issues in this area. There has been tremendous growth in the incidence of R&D collaborations and institutions that foster such collaborations. I'm referring to research joint ventures, strategic alliances and networks, licensing agreements among companies and between universities and firms, sponsored research agreements, industry consortia, and cooperative research and development agreements between firms and federal laboratories. In addition, there are what I would call property-based initiatives to foster collaboration, such as science parks, many of which are located at universities, incubators, and industry/university cooperative research centers. And then we have scientific collaboration, and studies have shown that scientific collaboration is increasing as well.

There are several factors that I think have led to the growth of these collaborations. The first is basically a change in the U.S. national innovation system to promote collaboration. This includes explicit policies to form public-private partnerships with respect to technology. The advanced technology program is an example of that. Policies promoting collaborative research, the National Cooperative Research Act, and the extension of that Act are examples of that. And then we have a whole range of policies – the Bayh-Dole Act and Stevenson-Wydler Act, among others -- to promote technology transfer among universities and federal laboratories and firms. Apart from public policy influences there is clearly an economic motivation for companies to outsource R&D and to share the cost of conducting R&D. Finally, the process of globalization has certainly stimulated international collaboration.

At the same time there is a burgeoning interdisciplinary academic literature on R&D collaborations. Economists, strategy professors, public policy, sociology, and financial economists are looking at different aspects of collaboration—the strategic implications, the economic implications, the financial implications, and even the social implications. Because of a lack of government data a lot of these researchers are relying on proprietary databases, and they are also going out and collecting their own data on partnerships and collaborative effort. Their analyses have been at various levels of aggregation, the individual scientist level, laboratory, the firm, the industry, or the university. These researchers use a wide variety of performance indicators, and they use a mix of both quantitative and qualitative methods.

The major data sets include the Merit database developed by John Hagedoorn and housed at the University of Maastricht. NSF has supported the development of two databases on collaboration. The CORE database which Al Link is responsible for, and the NCRA-RJV database, which Nick Vonortas is in charge of. But then we have again a long list of proprietary databases that have been used to study the different aspects of collaboration -- the Securities Data Company, ISI, Recombinant Capital, and so on. We also have data on university partnerships from the Association of University Technology Managers.

The critical need is for better information on the private and social returns to collaboration. Why? Because the rationale for government intervention is that there is a divergence between the private and social returns and so these programs and all of these various policies that I mentioned earlier are supposedly addressing some kind of market failure. We have no sense of whether those market failures are being alleviated or exacerbated even by the policies.

What are some of the performance indicators that researchers have looked at? They include patents, stock prices, total factor productivity, various measures of R&D, citations of patents and academic articles, licensing of technologies, co-authoring between academic and industry scientists, and the formation of start-ups, among others. The output of the collaborative effort is typically not directly observed. As a result, researchers rely on indicators of output and performance at the firm level.

I have some suggestions for research evaluation initiative that might help us estimate these private and social returns. First, I think it would be useful to ask some questions relating to the output and performance of these collaborative efforts. We need better information on technological diffusion and the creation of new products and firms and industries because that is often the result, we think, of these kinds of initiatives. Second, it would be useful not to reinvent the wheel, to work with existing data sets and organizations that have data on collaborations in an effort to generate better data for policy analysis and other kinds of research. Finally, it would also be useful to facilitate linkages between existing data sets on collaboration and linkages with data sets on economic performance such as the files that exist at the U.S. Census Bureau.

Our first speaker on this topic is Joshua Lerner from Harvard University.

JOSHUA LERNER: What I am going to say can be characterized as a series of footnotes on the issues that Don very cogently raised. First I'm going to talk about why strategic alliances and more generally these collaborative R&D forums are an important set of issues. Secondly, I'm going to address the problems with the data, particularly the public data, relating to these areas. Finally, I will suggest a couple of possible strategies for addressing these issues.

Why am I in a position to talk about this? I have myself done a fair amount of work on strategic alliances, for the most part not using typical public data sources but rather SEC filings and Recombinant Capital data sets. I also helped organize an NBER conference on strategic alliances last year where we tried to make progress on these issues.

Now why is this an important issue? When we look across a broad array of technology intensive industries, whether life sciences or information technology, we see a much greater reliance on alliances. These take several different forms. One form is pooling resources to fund cooperative research. But we also see a growth of what we might term as ex-post alliances, such as patent pools or standard setting organizations, where firms are sharing their intellectual property, the fruits of their R&D.

These activities seem to be playing a dominant role in many industries. As an example look at the flow of capital to new biotechnology firms. Not surprisingly, venture capital is very important, as are public offerings, either initial public offerings or seasoned equity offerings, the SEOs. These are all very important sources of finance, but the dominant source of finance is alliances with large firms. This was found by undergraduates plowing through SEC filings by biotech companies. The effort could be duplicated for telecommunications companies or other sectors. It highlights an incredibly important source of R&D activity.

Nevertheless, we have a lot of problems getting data on what is taking place. A significant number of agreements are being filed with the Securities and Exchange Commission but by no means are all of them. In particular, agreements between privately held companies will not be disclosed. The same is true of agreements between large publicly held companies where the investments and expected revenues are not large enough to be considered material to the companies' performance. Moreover, in many cases the information that is made available is lacking the details of greatest interest such as the dollar amount. And where firms issue press releases they may be misleading, for example in terms of the value of agreements. It is a bit like a drug bust where the police claim to have confiscated $10 million of cocaine. They are typically adding up all of the contingencies

The other side of the coin is that even when these alliances are disclosed, it is not generally announced when they are terminated. For those of us who sort of believe that the U.S. system of regulation and disclosure is robust it is a little disheartening to realize that alliances between biotech firms or internet firms often simply disappear. Only two years after being trumpeted as being fundamental for the company, it is not reported. Where did it go? When did it go? Why did it go? There's no clue in the public record as to what happened. So this is indeed a challenging area in which to collect reliable, useful data.

It is easy to conclude that what one is seeing in reported numbers is a tiny slice of what is happening. Consider the numbers in *Science and Engineering Indicators 2000*. According to the Merit database, which is compiled from press accounts, there were 574 alliances strategic alliances. Under the National Cooperative Research and Development Act, firms may file agreements with the Justice Department to enhance their antitrust immunity. According tot hat source there were 39 filings this year. Now look at one proprietary data source that Don mentioned, Recombinant Capital. They cover only two industries, information technology, somewhat narrowly defined, and biotechnology. They reported almost 2,000 separate biotechnology alliances and almost 5,000 information technology alliances involving at least one U.S. firm in 2000. So the data we are getting in a lot of these reports is really just the tip of the iceberg.

Another problem is that a lot of the information that would be most interesting for research and assessment is missing from the reports. The kind of things one would like to know are not easy to determine – the scope of the alliance, its size, how broad the area of activity, how important the alliance in corporate strategy, the structure, how exclusive or non-exclusive is the arrangement. And as I alluded to before, what is the duration of the agreement? Nor is the data structure as it stands capable of picking up the arrangements that emerge after R&D is completed.

An important question to consider in addressing these issues is who is going to be involved in collecting this data. It strikes me that today the best job of tracking is not being done by the federal government nor by academics but by the private sector. Not surprisingly, given the amount of money that companies are spending on business development, these firms are able to charge premium fees, which then allow them to invest considerable resources in data development and cleansing and scrubbing. Furthermore, organizations like SDC and Recombinant Capital have tried to go beyond simple counts of alliances, scorekeeping, and look at the nature of these agreements.

Finally, there is a need to go beyond ex ante alliances, agreements to undertake research, but also in the variety of cooperative arrangements that involve divvying up or using the research that has been done.

DON SIEGEL: Our next speaker is Nick Vonortas of George Washington University.
NICK VONORTAS: I have been working on this topic for almost 20 years now. I speak from the perspective of a practitioner, one who tries to collect and use such data There is no doubt that something has happened in the past 20-25 years. Nobody knows whether this is a fad or whether it will continue in the future, but clearly something is going on. In 1984 or 1985, I approached my Ph.D. thesis supervisor with a topic on alliances; I told him that I wanted to study research agreements. He looked at me in puzzlement: "What is that?" Stubborn as I am, I continued. Three years later I could not finish fast enough because the literature was coming from all sides. So clearly something is happening, and equally clearly we did not anticipate nor understand it very well. And part of the lack of developed theory is attributable to the lack of data.

It is not that we have not tried. NSF has played a very important role in helping the community develop data. They supported a meeting that I helped organize with Don, Al Link, and John Hagedoorn. We dealt with the policy needs for such indicators, we dealt with data that were available then, and we also tried to provide some initial advice to NSF on how to go about collecting data of this kind.

The results are reported in two publications, one already issued by NSF and one that is forthcoming in the Journal of Technology Alliance and Strategic Management, probably in May. One conclusion is that strategic research partnerships are important for emerging industries but also for mature industries and for industries that have changed structure. It is not that we don't have any data, we actually we have a lot of data. The previous two speakers talked about some of these. The SDC database, produced by Thompson Financial, is the result of collecting announcements and scanning all of the available popular literature. It supposedly covers everything, new agreements across all countries and all sectors. The second type of database, represented by CAPTI, is somewhat narrower. It captures strategic alliances with some kind of an explicit technology content, whether the conduct of research or the exchange of results, globally and across all sectors.

The third type of data gathering focuses on particular areas. An example is the Milano Polytechnic database on IT that was developed in the 1980s. Another example in the same field is the database compiled by Information Technology Strategic Alliance, ITSA, a very small company in California. The fourth type is represented by the work of Al Link. It involves a well defined group of alliances of one kind that is more homogeneous. In our case as defined by the Department of Justice.

A fifth type is currently under development at the George Washington University. This involves not just counts of alliances but linking data on alliances with participating company-level data on patents and other business information such as sales. One such database is being developed in the United States with NSF support. Another is being funded by the European Commission. I participate in both to try to ensure that the methodologies of the two databases are identical so that in the future we may undertake comparative work. We hope these data sets will enable us to go beyond previous ones in allowing us to see the network of alliances of a company or an industry and some of their outputs.

The data that we have currently, although voluminous, have clear limitations. Economists have done a good job of providing definitions. Alliances span a wide variety of things and our definitions are poor, in my view. We also lack analytical hypotheses. The business schools have tended to take over this field and they are not very interested in the same kinds of questions that interest policy makers. It is a major analytical problem that we can identify the beginning of alliances but seldom know the end. There is an issue of bias and idiosyncrasy. Each of us involved in this effort to collect data is interested in a somewhat different set of questions; this influences how we collect data. One way to mitigate this is for NSF or some other organization to push for consistency and especially for linkages to data on all the other things that companies do. The Europeans have approached this through the Community Innovation Survey (CIS), which has been conducted three times The survey asks a significant sample of companies where they ask not only about alliances but also about strategies for new product development and for research more generally. NSF has attempted a pilot innovation survey in the United States, but it is not clear whether that will be repeated or expanded.

If you look at alliances a little bit more closely, they are in three broadly defined fields -- advanced materials, information technologies, and biotechnology. These are so-called infrastructural technologies used in other sectors. Biotech is useful for pharmaceuticals, agriculture and other applications. To understand alliances we need to understand the uses of these three technologies in the economy.

There is clearly a need for multiple measures of inputs and outputs, because the nature of alliances is so diverse that we cannot capture it with one or two. And I am increasingly convinced that we need a better understanding of organizational structure. One alliance may be an equity based agreement, another may be a contractual agreement with 15 different kinds of underlying outputs, and a third may be simply a one-time licensing agreement. It does not make sense simply to add them up.

JOHN ALIC: Nick alluded briefly to the automobile business, which is very interesting because it's the biggest industrial R&D performer in the United States, yet almost all that R&D is actually product development. That industry is shifting towards you might call collaborative engineering between the OEM's and suppliers. And you didn't mention that at all. Is that because it is outside again what you mean by strategic alliances and if so, why?

NICK VONORTAS: It is not outside of what I mean by alliances, but simply because we know it is important does not mean that we can obtain data on it. An arrangement between Ford and a supplier may not be reported in the press. If it were, we would pick it up.

BILL LONG: Give me a number for the percentage of R&D scientists and engineers engaged in alliances relative to the old model, or R&D dollars, or patents, or any measure. Are we talking about 1 percent, or are we talking about 21 percent?

BRONWYN HALL: I'm guessing that it is somewhere between 5 and 10 percent and probably closer to 5 percent. With regard to patents there is such a thing as a patent held by two entities but it is extremely rare because one of the first things they do when they sign the agreement is figure out how to allocate intellectual property rights and who's going to own the patents. So you can't use the patent data to do it.

JOHN JANKOWSKI: I cannot give you the number. But two questions have been added to the industrial &D survey form within the past year. One is the amount of collaborative R&D, how much of the R&D that's being reported is used in collaborative R&D activities with for profit companies, universities, non-profits, so that is something that is beginning to be collected. Another question that was added has to do with R&D of an infrastructural nature – what is the level of effort in biotechnology, materials, and software. This question is not tied to alliances so we won't have a clear idea of the proportion but we can begin to make a stab at answering the question "how much relative to the whole?" Researchers able to access the micro data in theory will be able to look at the firms that have a mix of activities and go further downstream.

PARTICIPANT: Based on activity in our business one would have to say that alliances are increasing. But are they long-lived? Do we have any information on their duration?

NICK VONORTAS: This is one of the big unknowns. This is not reported, and in fact no firm will find it useful to announce to the newspapers that it is giving up on its alliance with Y, so we won't know, unfortunately.

CHARLES DUKE: Alliances are made to complete value chains, and value chains are offering specific things or a particular product or a particular service. Xerox has a huge variety of these alliances and generally speaking they last a year or two at most. They are changing constantly, and as the product line changes, the alliance network changes. Now there are certain other kinds of alliances that are enduring. We have a relationship with Fuji Xerox that has existed for 30 some odd years and that contributes something like one-quarter of our earnings. So there is such a thing as a permanent alliance based upon marketing arrangements, and who has access to what markets and who provides what technology. So the paleotological notion that you can look at the bones and figure out where the animal went is not a very useful notion. The only way to make sense out of it is to understand what these firms are trying to do, what their product line is, whether they have holes in their value chains, and whether they are trying to fill a hole in their value chain by making an alliance

PARTICIPANT: Perhaps the notion of trying to determine the end point of an alliance is unrealistic because they don't actually end, they just take different forms, they evolve, and you can see this in work we've been looking at in the engineering research center, which gets funding from NSF for a period of time and then the money goes away, but the centers don't go away, and the vibrations don't go away. They tend to change dimension, they change significance, the free riders disappear because they no longer can benefit. And I suspect that is true in the industrial side.

LOCAL R&D ACTIVITY

PATRICK WINDHAM: This is a panel on the issue of location of R&D, both domestic and international. A few years ago there was the buzz phrase of "think globally, act locally," and while that phrase has been overused it is striking to me how the policy community that uses R&D data has expanded. For example, on the domestic side a few years ago I was working with a group in San Diego, and we asked what we thought at first was a simple question. How much federally funded and privately funded R&D is done in the county of San Diego? It turned out to be nearly impossible to get those data. NSF does a fine job on state level data and is paying more attention to the sub-state level. But a few years ago as the cluster theory of regional economic development became popular and local business and government leaders were wondering how to build their own clusters, we found out that the data were not easily translatable. One of my colleagues, Caroline Lee from UC San Diego, worked with the RAND Radius Database, and that was how our team was able to put together a good picture of public R&D in San Diego, but it was a considerable effort. We are going to talk about the value of R&D data that are broken down in both sub state regions and also in metropolitan areas that cross state lines.

There is also the issue of cross national data. What are other companies based in other countries investing in the United States? What is happening with U.S. companies investing overseas? In California the big question is how much engineering design and R&D is moving to China. Not just production. We are wondering how big that trend is and what is the long term impact. Our leadoff speaker is Rob Atkinson of the Progressive Policy Institute (PPI) who has conducted several studies comparing the economic performance of metropolitan areas, regions, and states.

ROB ATKINSON: Before I joined the PPI, I got into the issue of R&D at the state and local levels as director of the Rhode Island Economic Policy Council, which was a state initiated public-private partnership to address the future of that state's economy.

In Washington it is easy to forget that there are a lot of other governments in this country that are intently focused on R&D policy. You may think that it is solely federal agencies. I recall being in a meeting a few years ago with the head of DARPA who said "R&D doesn't take place at the local level; it's only inter-firm collaboration that's important, local collaboration has no meaning." This was when we were developing the Technology Re-investment Program (TRP). Since then awareness has increased at the federal level in terms of activity at the state and even the metropolitan levels. But I think one of the major things for people who are working at that level, one of their major frustrations in life is the relative lack of good federal data. Often times they are just shooting in the dark.

It matters because 10 years ago most of what states and metropolitan areas did was chase smokestack industries or branch plants; but in the past 4 or 5 years the economic development strategy has changed dramatically toward capturing value from the new economy. Take a look at a website www.SSTI.org of the State Science and Technology Institute (SSTI). They have a feature called "tech talking governors," which highlights the statements on technology policy in the 50 state-of-the-state addresses. This year they are talking about technology and the knowledge economy.

A report we did called the 2002 State New Economy Index (an update of one we did one in 1999) attempted to sell states on the fact that there is a new economic framework that they need to focus on. It is not about cutting costs anymore nor about having a "good business climate." It is about having a good innovation climate. That will determine your success as a state or a metro area. One of the indicators is, of course, R&D expenditure. We rank all 50 states in terms of RD intensity. In fact we did something interesting this year. Controlling for their industrial make-up, we ranked states on R&D intensity. The number one state was Rhode Island. You may think I manipulated the variables to get that result but I did not. Number two was Idaho, the second most R&D intensive state in the nation when you control for industrial mix. So the results were interesting.

Why is that all important? It is important because more and more governors, state legislators, and economic development directors are getting the message that they have to do something. But often times they are shooting in the dark. Another reason data are important is the emergence of a variety of state and metro-level economic guidance organizations, like the one I ran in Rhode Island, whose main mission is to analyze in some detail the local economy. It is crucial for them to have good information. Furthermore, in just about every state there is now some kind of technology promotion organization, the Ben Franklin Program in Pennsylvania or the Oklahoma Science and Technology Authority, whose main mission is to promote technology, led economic development. Good data are also important to them.

The final reason good state and local data are important has to do with what we want to be happening from a national perspective We want states to be in intense competition with each other, but not to cut taxes on interstate company location. We want them to be in competition with each other on who is going to invest the most in these kinds of activities. And one good way to get them to compete with each other is to good data so they can compare themselves to each other. As long as there is competition to do more R&D and more training, we should want to encourage it by ensuring that there are good data.

What are the major problems with local data on R&D? By and large it doesn't exist at the metropolitan level. Although there are sometimes confidentiality problems at the state level, in most cases you can get aggregate state level data but rarely but rarely any real detail. NSF has some detail for the ten largest states, as I understand it, but for most states only one number. It tells you something but not nearly enough to help in developing micro level policies or in persuading a governor that the state has a big opportunity in, say, cellular medicine. You have to rely on anecdote or talking to individual companies.

The last problem with federal data is its age. We now have 2000 data but and here we are almost mid-way through 2003. Time lags of this magnitude are significant for states, particularly in a rapidly changing economy. Some states are beginning to question the "new economy" represents the right direction. All you can say is that we have data for the boom years. So I think the issue of timeliness is a central one. Technically it seems like a straightforward issue to address although it may mean spending more money to produce data faster.

How can we deal with the issue of confidentiality to avoid disclosing individual corporate behavior? I think there are several ways it could be done. One is to break the data down into some broad categories for smaller states to give some granularity to the data. One breakdown might be basic research, applied research, and development. Another might be company size -- R&D in big companies and small companies. You might divide it into durable and non-durable goods and services–three categories. You might attempt to figure out how much is biotech, how much is IT, and how much is materials. But some level of detail possible without violating company confidentiality. It would help to report on the top ten metropolitan areas as NSF does for the top ten states. We could get some real sense of how much New York is doing versus LA versus Seattle versus Santa Clara County and San Francisco.

An example of progress is a new report for the Economic Development Administration on metropolitan technology transfer by Andy Reamer, a consultant in Massachusetts. He put together an entire R&D database at the metro level, so it took him forever. Anyway, I encourage you to look at that report because I think it represents the state-of-the-art. When I asked him what he would fix in the federal data portfolio he said there is a need to add data on nonprofit institution spending, federal funding of non-profit research institutes at the metro level, intramural federal R&D, and certainly this industry R&D I just mentioned.

My last point is that there is a problem with where R&D is assigned and where it is carried out. Rhode Island was fortunate to have the Naval Undersea Warfare Center whose R&D spending pushed the state to the top of the chart. But when you talk to people in some of these installations they say we don't really do it here, we do it somewhere else. Thanks, but that does not help me much. Getting a better handle on actual location of performance as distinct from place of reporting is also central.

INTERNATIONAL R&D ACTIVITY

PAT WINDHAM: We are now going to switch to the international side and are pleased to have Walter Kuemmerle from the Harvard Business School to begin our discussion.
WALTER KUEMMERLE: I think my best use of my time is to give you a complementary perspective on data sources by describing my research providing a perspective on the globalization of R&D. The research also presents an alternative taxonomy for classifying R&D in geographic space. The whole point of what I'm talking about is that location matters, and that we do not understand very well how local spillovers happen although they are extremely important, be it in Rhode Island or in Boston or in Beijing.

What I am going to talk about is primarily micro level data. One of the advantages of being at a business school, especially when you have some resources, is that you can go out into the field and try to understand what people actually do. This is not a substitute for but an essential complement to the structural data we get from the Census, BEA, and NSF to get a sense for how things really work.

I also want to comment on the memorandum of understanding concerning sharing of Census data, BEA data, and NSF data. I think this is a great advance. On an international level, we know very little. But although we complain about inadequate data, by almost any comparison with other countries, even advanced countries, the quality of data available here in the United States is actually quite good.

So with that as a starting point, let me describe the study I did on foreign direct investment in R&D. I was motivated to do this by the fact that I saw that there was an enormous increase in R&D being done by companies across borders. The prior studies that had been carried out on this subject at the firm level generally just involved one source country or one target country. They had very small samples and they were not across time. At first you hope there is some data set that you can mine by running some sophisticated regressions. But that is not at all the case here. It was necessary to go into the field and collect it.

What I decided to do was to select 10 U.S. companies, 10 Japanese companies, and 10 European companies in two industries that are technologically intensive -- pharmaceuticals and electronics, specifically computer hardware -- and to do a very detailed data gathering effort by assuring these companies confidentiality and providing them with a questionnaire that they would not answer if they simply received it in the mail.. They only participated if the CEO committed to the study and instructed staff to complete the questionnaire. Even if companies are required by law to participate in a federal survey there are ways to avoid spending the effort to provide absolutely accurate answers.

What are the findings of this study? The first finding is that you can think of R&D in a different way in space, and this is not just across borders, it is also within borders. If you think of the home base as the place where a company has its core competence, local activities augment the home base by taking spillovers from some local environment and transferring that back. The alternative to doing R&D locally is obviously licensing in, but when you license in you license some very specific intellectual property. What you want to do when you build an R&D site is to seek to capture some spillovers although you don't know precisely what they will be. This is why you go there to carry out R&D. This implies that you need a significant minimum local presence, so you are taken seriously in the local community. A mistake that a lot of Japanese firms made in the early 1990s was that they would set up these listening posts that didn't provide them with a lot of insights because they were not taken seriously locally in the communities they entered.

On the other hand there was a lot of home based exploiting activity going on -- activity that makes use of whatever is available at the home base in a more efficient manner, either for R&D or manufacturing or marketing. In Japan, for example, a pharmaceutical company is required to rerun all clinical trials locally on the premise that physiology is somehow different. That requires that you set up local R&D units.

So one alternative taxonomy is home base augmenting versus home base exploiting. In terms of location, as you would expect home based augmenting facilities are built close to universities where the spillovers that you are seeking are highest. Home base exploiting facilities are generally close to centers of demand or close to existing manufacturing facilities.

The inter-temporal characteristics of R&D location are important. I collected all this data for all the companies from whenever they started doing R&D abroad. There is an increasing number of home based augmenting facilities that get built, which suggests that even companies that are very, very powerful in terms of their innovative capacity sense a need to go elsewhere to bring technology in-house that they think is useful for them in the long run. A lot of this appears to be greenfield development because the risks associated with acquiring something that you don't understand are too high.

Of the 30 companies in the sample 20 are Fortune Global 50 companies, even though the

Japanese pharmaceutical companies are relatively small because Japan at that point in time was somewhat weak in pharmaceuticals. Whenever a firm declined to cooperate, I just went to the next largest company. According to my calculations, the sample captures actually a significant share of R&D worldwide in these two industries.

There is an article in the *Research Policy* (1999) that summarizes some of these findings in more detail and another one in the *Journal of International Business Studies* that looks at this taxonomy. The firms in the sample established a total of 155 R&D sites abroad, an average of five per firm. This is a substantial managerial challenge for a firm. How to ensure that the relationship between the laboratory in Grenoble and Xerox's home base works well is a big challenge.

From 1975 let's go forward to 1985 and then 1995. The number of foreign locations has increased. Second, the relative share of home base augmenting facilities is increasing. This is a game largely being played amongst industrialized countries, with the exception of China where you see some home base exploiting facilities, and India and Southeast Asia, where you also see some home base exploiting activities. Developing countries, at least for these two industries, are not on the map. If you included motor vehicles you would find some home based exploiting activities in South America -- exploiting, not augmenting --, because innovation in motor vehicles is primarily done at the supplier level, and suppliers are located in industrialized countries.

This study was quite resource intensive and probably could not be replicated for a much larger sample, because it really required me to visit these firms and talk to at least three executives there, sometimes more than once to clarify issues and get additional data. But it is interesting that such a study simply didn't exist at that point. Available data that I examined and sometimes used for other papers existed at the country level and some at the state or county level even though I am not at all surprised that it took an enormous effort to get this for San Diego County. There was of course data at the industry level and data at the firm level but there was very little data at the establishment level, which is needed to understand spillovers and design policies for these spillovers. :Rob Atkinson's Naval Research Center is a great example. If you don't know what an establishment really does then you cannot design effective policies for it.

The private data collected by consulting organizations, as Josh Lerner indicated, is often quite good, but it is generally very narrowly focused. A less good example is the directory of American research and technology which is published every year primarily for vendors who sell to R&D labs. The data on R&D lab size are simply inaccurate. They list the number of employees; it is not very reliable but it gives you some indication.

It is definitely desirable to have more detailed data on R&D -- location, activity level, and focus are very important. But doing this at an internationally comparable level will be an enormous challenge because other countries, including quite a few industrialized countries, have a political problem providing the resources necessary to collect data at the level of detail that we need.

I agree that it is difficult to classify R&D data at the sub-company level simply because companies are very differently structured. A group that could provide useful advice on this issue is management scholars who study industrial organization. They might suggest ways of collecting such data consistently although they often differ on how firms should be organized.

Finally, I think providing useful information back to respondents is an important way to encourage cooperation. There is an opportunity to do a much better marketing job for data collection from the private sector.

MR. WINDHAM: The Bureau of Economic Analysis has done a lot of work in this area, and we are pleased to have Ned Howenstine here.

NED HOWENSTINE: I am in the international investment division at the Bureau of Economic Analysis, and unlike the rest of the Bureau, my division actually collects data that we produce. The rest of the Bureau basically is a user of data from the other statistical agencies in the federal government, but I am here as a data producer.

Our programs mention that there is some international data on R&D available from the current NSF-Census RD-1 survey. I'm here to discuss some other international data that my agency collects on multinational company (MNC) R&D activity. This may not be quite as familiar to this audience as the RD-1 data. I also want to talk about a new proposed project that we hope will improve the international R&D data.

BEA collects data for both U.S. MNCs and foreign MNCs in annual and benchmark surveys. We collect U.S. MNC data in our surveys of U.S. direct investment abroad, and in those surveys we collect data on U.S. parent activity in the United States and on foreign affiliates' activity abroad. For foreign direct investment in the United States we collect data on the U.S. activities of foreign MNCs. Our benchmark surveys are carried out every 5 years. They are comprehensive, covering the universe of foreign direct investment in the United States and U.S. direct investment abroad. In the years that we don't do a benchmark survey we do an annual survey, which is a sample survey that collects less data. Both types of surveys are mandatory and conducted on an enterprise basis. Of course our authority does not extend to the foreign operations of the foreign MNCs.

Annually we collect total R&D spending. In our benchmarks we have some additional detail -- for whom the R&D is performed, whether it is performed by the parent or affiliate for themselves or for others, and in the case of U.S. parent companies whether it is performed for the U.S. government. We also have some information on R&D performed for the affiliates and parents by others outside the organization, and in the benchmark we get data on R&D employment. For foreign MNCs we have annual total spending on a performance basis and total R&D employment. And in the benchmark we have additional information on for whom the R&D is performed and on R&D performed for others for the affiliate.

Our most recent data are for 2000, admittedly a bit out of date. U.S. parent companies in 2000 spent about $132 billion dollars on R&D, representing about 66 percent of total U.S. R&D spending. That share is high relative to the U.S. MNCs' share of GDP, which is about 21 percent. The 66 percent share, by the way; it was 77 percent in 1994. The decline does not appear to be the result of parent companies' shifting R&D abroad. Foreign affiliates spent $20 billion in R&D in 2000. If you add those two numbers together you get about 87 percent, of total U.S, R&D, about the same share as in 1994.

Turning to foreign MNCs, they spent about $26 billion in the United States in 2000, about 13 percent of the U.S. total. That share has increased slightly, from about 11 percent as a consequence of the surge of foreign investment in the United States over the past few years.

I just need to point out some overlap in the U.S. parent and the U.S. affiliate data. A U.S. parent company can be a U.S. affiliate. If Honda buys General Motors, General Motors becomes a U.S. affiliate of a foreign company. But because General Motors has its own foreign operations or foreign affiliates, it is also a U.S. parent company. In our R&D data, there is about a 14 percent overlap.

There are some drawbacks with regard to the BEA data including questions about their comparability to the data produced by NSF. Partly because of these drawbacks we have a joint proposal to try to enhance the international R&D data available. Specifically, we would attempt to link our data that we collect in the direct investment surveys to the data from the RD-1. At this point it is simply a feasibility study. It will cover our most recent benchmark survey years which were 1999 for U.S. direct investment abroad and 1997 for foreign direct investment in the United States.

The project is being sponsored by the NSF. It was prompted by an NSF request to add some questions to our 2002 survey of inward foreign investment in the United States. We realized that those questions would impose an additional burden on respondents and further that some of the information being sought is available in the RD-1. It occurred to us to try to link our surveys together and avoid additional costs to the companies that report to us.

If successful, we hope to update the results annually; and the linkage could be extended to other data sets. This could be done under two legal authorities, a 1990 act that allows BEA to link its data on foreign direct investment in the United States to establishment data collected in the economic censuses and a new 2002 statute on data sharing

We are going to attempt the linkage first by doing a computer match of employer identification numbers that are reported to both BEA and Census. If we are not successful on the computer match then we will go to other information such as names and addresses. The first report of the project by our friends at the Census Bureau will discuss how successful we were in actually linking the data. It will talk about the numbers of tables that we think we can produce, and the types of tables that we think we can produce. Of course, here, confidentiality is a major consideration. It will discuss the feasibility of moving the link forward in time and if we think it's feasible we hope to produce the methodology for doing that. Finally we hope to produce a number of tabulations on the data that we get out of the project.

We hope eventually to have a better understanding of the international features of R&D, including the dimension of ownership and the funding. Specifically, for operations in the U.S., for U.S. parent companies and U.S. affiliates, we will have the core data that is available from the RD-1, such as a number of R&D performing companies, total R&D spending, R&D employment, sales of R&D performing companies, total employment of R&D performing companies, and state location of R&D activity. We should be able to tabulate this data by industry and for the U.S. affiliate data we will be able to tabulate it by country of foreign owner. For the overseas activities of U.S. MNCs we will be able to get counts of number of parent companies with R&D performing affiliates and various data for the R&D affiliates themselves -- R&D spending, R&D employment, and sales of R&D performing companies and employment of R&D performing companies. We should be able to do tabulations by industry and by country of foreign affiliate.

In addition to the core items we should be able to get various types of data that are available in the RD-1 that BEA doesn't currently collect for MNCs such as biotechnology research. We do have some source of funding data in our surveys but not as detailed as in the RD-1, again by U.S. location, type of expense, and type of organization for non-company R&D performers. None of this data is currently available.

By doing this matching exercise we expect to improve the quality of our data and the NSF-Census Bureau data. We should be able to uncover erroneous or missing data, issues of industry classification, and issues affecting reporting. There are issues of definition -- ours follow the NSF definitions fairly closely but not exactly – and consolidation. At what level of consolidation do enterprises report on each of the surveys?

BEA surveys are conducted on a fiscal year basis. We don't think that that has a very significant impact on the data because most of our respondent companies' fiscal years end at the end of the calendar year, but by comparing our results to those of the RD-1, which is on a calendar year basis, we will learn more about that question.

We sample in our annual surveys so we have concerns about whether we might be missing R&D data in some industries or countries, so we will learn more about the effects of sampling on coverage. We expect to be able to improve both our sample frame and the Census Bureau sample frame. We expect to find cases where companies that are reporting to us are not reporting to the Census Bureau even though they should be reporting, and vice versa.

PATRICK WINDHAM: For our final speaker we are fortunate to have Robert McGuckin, senior economist at The Conference Board.

ROBERT MCGUCKIN: I first want to commend the NSF-Census and BEA data linking initiative. It is a great idea. In the years I was at the Census Bureau's Center for Economic Studies we did a lot of linking of surveys and concluded that it generally yielded both methodological and analytical benefits. This project seems to me a good one to start with.

I am going to discuss a work in progress, so these are not final conclusions but initial impressions. We anticipate a report in October. We are coming at the international R&D question in two ways. First, we are conducting very detailed company interviews. Second, we are developing R&D purchasing power parities (PPPs) to enable more accurate comparisons across national currencies. We are doing this at the country and industry level. We started out with 1997 but we soon pushed back to 1987, ensnaring us in some serious statistical issues dealing with PPPs done with different methodologies over time.

We have completed interviews with 25 R&D executives, typically CTOs, and we will probably do 10 more interviews in four industries -- drugs, telecommunications, computers, and autos. We have also conducted five interviews outside those sectors as benchmarks. In the case of a few companies we spoke to both U.S. and European executives, for example IBM and Ford.

First a couple of observations. R&D are quite different activities with different drivers and different inputs, production functions, and outcomes with different uncertainties. Second, we are observing enormous changes in business structures and management technology. Organizations are seeking to improve the productivity of R&D. Basic research is a non-starter for the business firm. They simply aren't doing it anymore. The day of the regulated monopoly is gone. If they need basic research they do it through an alliance with a university or individual professor.

PPP adjustments are very important. I'm often amused when I read that Japan is the world's second largest economy. If you look at adjusted PPPs China and Japan are far behind. And it is also true in our numbers; PPPs make a large difference.

Applied research generally entails some considerable technical risk and uncertainty about application. If I go to the development side the intended commercial application is generally known. There's a fuzzy line here between this research and development, you can call it early stage development or applied research it is hard to draw that line. But market risk is the large risk in development, not risk about whether you will achieve a useful application. Also the time horizon once you get to the development stage is generally much lower – perhaps three years but two years on average. It is a very short process. The scale of resources required is also quite different. Most development is much more material intensive.

The key organizational change that emerged in almost all of the interviews was the shifting of development to the business unit, to the operations unit. This reflected a focus on a commercial objective and the need to have the marketing people involved as well as take account of the manufacturing process.

So what are the implications? We expect these organizational changes to continue and proceed rapidly. That has to influence how we deal with data collection – the importance of the business unit, the emergence of matrix management, the involvement of non-R&D executives in decision making, the shift of resources from the research unit downstream, and the growth of outsourcing.

The question that is hard to get a handle on is what is happening to productivity. If we think these organizational changes are productivity enhancing, which is my bias from studying other kinds of organization, we may be seeing some real improvements in R&D. And if information technology is driving research as it has been driving other parts of the business enterprise, we are seeing enormous productivity benefits. But the work so far is all in manufacturing firms so we need to study the service sector.

Every business said that cost doesn't matter very much as a factor in location. But in the United States the relative costs have been declining and we are seeing R&D shifting here. It has been shifting away from Japan and Germany and France where the cost is higher into Britain and into China and India, although the latter is more development than research. So cost may be an important factor and that is one of the things that we are trying to determine. Another factor in the expansion of U.S. R&D is that business has been expanding and with it the overall intensity of R&D.

With research there is a clear home country bias at the level of the centralized research laboratory. It also has to do with the proximity of universities and non-corporate centers of research capability. A lot of companies are coming into the United States to access computer technology, especially in Silicon Valley. We found that U.S. firms' acquisition of foreign research laboratories is generally associated with mergers. Companies pick up a laboratory in connection with a big merger and decide to integrate it.

The location of development activities has a lot to do with local regulations, proximity to marketing expertise and to the suppliers for that marketing operation and manufacturing capacity. There are country-specific standards in telecommunications, local regulation of drug marketing, and safety regulations in every country. Often there is a preference of local models of computers and automobiles. These factors tend to run in the same direction as cost differentials, which are larger if you use R&D PPPs rather than the conventional GDP PPPs. Whether they are big enough to make a difference in policy remains to be seen.

This is a brief overview of what we are being told by executives about R&D performance.

ANDREW WYCKOFF: First a comment. The problem of timeliness that you face in the United States is not much different that some of the countries we are dealing with at the OECD. There may be some tricks that you can pick up from the Swedes and Finns. One advantage they have is that their economies are dominated by a few large firms, and I think they pay attention to the annual reports and you get a fair amount of information out of them.

I wanted to ask a question. When we look at multinational enterprises and their global reach, one of the striking findings of BEA is the amount of intra-firm trade. I am wondering if you have a handle on the amount of intra-firm technology flow, which could theoretically be picked up in royalties or licenses going back and forth. .

NED HOWENSTINE: We do of course have data that shows up in the balance of payments on international transactions and royalties and fees. I think the problem is that charges for technology transfers are probably not always explicit and may be embodied in products and in transfers of science and engineering personnel abroad to help with the R&D, transfers that are not measurable in the sort of aggregate statistics that we collect.

ANDREW WYCKOFF: I read in a *Wall Street Journal* article that some of these patent portfolios were actually kept offshore as well, and I wonder what that does to your country location figures.

NED HOWENSTINE: That is a problem. We're concerned about inversions and related R&D activity, and we're actually doing some studies of that ourselves right now. I'm not sure what the effect is. My colleague here Obie Whichard who specializes in services, and those kinds of transactions may have some insights.

OBIE WHICHARD: It occurs to me that sometimes companies are required to allocate certain charges for tax purposes and that may be what is reflected.

PARTICIPANT: There has a been lot of research on the tax figure; you might take a look at it because, at least until fairly recently, firms had quite a bit of discretion on whether they repatriated the research and the R&D, and that means using annual figures of any kind of transfer back to the United States. It's not informative because you can affect your taxable revenue streams simply by where you move it.

WALTER KUEMMERLE: I don't think that firms generally make strategic decisions about this for tax purposes. This is managed.

CHARLES DUKE: There are many different arrangements when you purchase R&D or when you have corporate organizations that have R&D, and the notion that a firm does one thing is a flawed notion. These arrangements are very specific to the value chains of the product lines involved, even within a firm.

NED HOWENSTINE: The RD-1 has a question about total foreign spending for R&D without distinguishing whether it is R&D by an affiliated company or not. By combining that R&D RD-1 data with our data about affiliated relationships we might be able to learn a bit more about the MNC activities with regard to arms length versus affiliated R&D spending.

BILL LONG: If state governments are promoting technology in order to have economic growth, why in 2003 are they slashing technology supports along with other programs?

WALTER KUEMMERLE: The budget crisis will have various effects, New Jersey just eliminated or dramatically downsized their efforts, but other states such as Michigan are talking about increasing theirs even in the midst of budget deficits. I think we will see a net reduction in state R&D investments, but I think the real question is just how long is the state fiscal problem going to last. Regardless, some states believe they have to keep up their investments for competitive reasons.

PARTICIPANT: I have heard that state spending on research, particularly industrial related research is cyclical. When times are good spending is up, when times are bad spending is down, which is exactly what you don't want to have happen. In contrast, federal spending of the same sort, limited though, tends either to be constant or actually a little bit counter-cyclical. It seems to me that timely data on states' own spending is genuinely important.

WALTER KUEMMERLE: NSF doesn't collect it because of funding limitations. The closest we have is a six or seven year-old study by State Science and Technology Institute (SSTI). I guess it depends on what state you're in and how bad their budget crisis is, but I can point out a number of states that are actually increasing funding right now, not cutting it. States are not going to go back to where they were. They have more or less embraced this new innovation driven model, although it will fluctuate according to budget cycles.

JOHN JANKOWSKI: NSF actually funded SSTI survey and how often to sponsor such a study is something on the agenda of the Academy committee to give us their advice. It should be an activity that we undertake on a more frequent or consistent basis.

PARTICIPANT: This is more of a comment than a question and it follows from what you have just been discussing. In the absence of really good data at the MSA level we have regions coming to us on practically a weekly basis asking how to replicate San Diego in their neck of the woods. Governors and mayors and regional authorities all want to do biotechnology in their backyards. It is my contention that you cannot grow biotechnology in everybody's backyard, but absent good data from NSF, regions are really in the dark. If they had better measures they would have less difficulty justifying their strategies.

CONCLUDING OBSERVATIONS

BRONWYN HALL: Before we conclude I want to recognize Al Johnson from Corning and from the Industrial Research Institute.

AL JOHNSON: We have a strong interest in improving the reporting of industrial R&D spending. What sort of action does that mean? What level of disaggregation? I am not entirely sure, but at the next meeting of the IRI Finance Directors' Network we will try to get opinions out onto the table concerning this issue and generate some options for moving forward in a cooperative way to provide more useful data for private and government planning while ensuring that companies' proprietary data is not exposed to their competitive disadvantage.

Bob McGuckin mentioned that companies don't do fundamental research. That's not quite true. What I've observed in participating in the IRI is that U.S. companies that do work in materials still do a substantial amount of fundamental research. Is the sample biased? I don't know, but I do know from personal experience that a number of companies are doing fundamental R&D.

What kinds of R&D data are now generated in companies? Well, people track the kind of data that you would expect for financial and managerial accounting. This is not a secret. The problem is that by rights different organizations organize that data differently; their taxonomies differ. So marching in and declaring to them that you want data in one way shape or form might not get you the results that you're anticipating. Not because they're trying to dodge you, although they may be, but because they stack it up differently in house.

If an organization manages its R&D portfolio using a stage gate process – meaning that there is an inception and some design and engineering and some prototyping and so forth moving through toward production -- they may in fact track spending by project stage, and the disaggregation may in fact get quite difficult.

If the Census asked for further classification should the level be line of business or some other category? The answer is I don't know, but my subcommittee chairs have agreed to talk about it to work with you toward a productive result.

BRONWYN HALL: The next speaker is Fred Gault from Statistics Canada, a member of the CNSTAT study panel.

FRED GAULT: Something I have learned today is that the R&D enterprise does not operate in isolation. What is important is not just measuring R&D expenditure but understanding the sources of funding, where that money to do R&D comes from in a firm, whether it's coming from government, the firm itself, or from other firms, perhaps classified to other industries, or whether it's coming from abroad. Now we heard about the memorandum of understanding to deal with foreign affiliates and the flow of funds, and I think that is an important step forward. What it doesn't address, but John tells me is in hand, is foreign payments for the performance of R&D in U.S. firms. That is a number on which you cannot lay your hands at the moment. In Canada, one-third of business enterprise R&D comes from abroad, so that makes it a very interesting number because if it disappeared it would have implications for highly qualified human resources in the R&D enterprise. So those linkages are important, and also the payments that firms make for R&D to other firms, to private nonprofit organizations, even to governments. People rent time on wind tunnels and other interesting objects as part of doing their business. And when you trace those linkages you begin to build a picture of the R&D enterprise.

It applies also to federal spending on R&D, grants and contracts, extramural spending, therefore and internal spending. Now people like our colleagues in RAND can take those numbers and produce state or sub state results, which I'm sure are both timely and useful. We produce them in Canada and the first thing that happens when those numbers are released is that people complain about not getting a fair share of federal spending and they want to know why. So there ensues a debate in the house, which mercifully goes away after a while. .

The unit of observation is clearly important. It is a firm level unit of observation for the RD-1, and that works for small firms, but it raises complications for large firms. We've tried to struggle with it and it is very difficult. We have a lot of R&D in wholesale trade, a large amount of it is pharmaceuticals, some is software, and you could argue that it is misclassified. Well, that's where those firms happen to land when they come out of the business register and following my rules and guidelines I have to put them there. So there will be for some time to come a lot of R&D in wholesale trade, but watch it carefully.

Timeliness. Well everybody wants the data immediately because they want to do their thing, whatever their thing is. And that raises an interesting question. There are ways of improving timeliness. In business R&D you could, because of the high concentration of research and development here and everywhere else, survey the top 50 firms and you could pick a sort of reduced set of variables that you could use in that survey in order to get the results out within a month or so of the reference period. R&D tends to be an annual thing because that's the way people do budgets. But before you contemplate that, I suggest sitting down with the policy makers to ask them exactly what they're going to do with these numbers and which variables are important to them and how much they're willing to pay for the accelerated service. That will focus the discussion a good deal, but it would also focus the work of the data takers.

From long and bitter experience it is my view that if you don't have the policy makers with their checkbooks sitting at the table, it is not a meaningful dialogue because they have to pay for the numbers, and once they've paid for them they will use them because they have to justify that expenditure. A meeting like this is amusing. We sit around and say that we would like more of this and more of that but we are not paying for it.

Outcomes of R&D are absolutely important. Notice I haven't used the word innovation yet. And we have talked a lot about value chain and understanding how R&D fits into the value chain, but not until a recent session did I hear a discussion of size. Let me tell you a story. We do innovation surveying in Canada. One thing that preoccupies us is the source of ideas for innovation. One item on that list is government laboratories and government programs, but virtually no one uses government laboratories or government programs as an input to innovation. Now how depressing for the minister, but re-cut the data and look at large firms. If you look at large firms they all use government laboratories and government programs, because they all have

R&D facilities and they all have the capacity to sock every bit of intellectual property out of everything that they can find in the country, inside or outside the country. So size matters and a lot of what we were talking about in our value chain discussions apply to large firms. And there are policy differences with respect to how you deal with large firms or small firms by way of tax credits or tax benefits.

That brings me to the issue of links, which has been a theme of our discussion. Intellectual property commercialization from government labs and, from university labs have been the subjects of very good surveys. We could make a greater effort to understand how government R&D gets out into small firms or gets licensed to large firms or moves around in the system. Other topics deserving more attention include alliances and networks and partnerships.

There are outcomes to all of these activities. We do the R&D. The R&D has some sort of impact. It gets commercialized as we move along the value chain, and then things change. Perhaps the labor force in the firm has to be upgraded to deal with the new activity or downgraded or replaced, who knows, but there will be social and economic outcomes. Profits will change, market share will change. That is something that is difficult to see until you have good data linkage back to the MOU.

We mentioned clusters. It is very hard for statistical offices to produce data at the level of a metropolitan center, very difficult indeed. We are experimenting with it but we have confidentiality problems. We would publish tables for a city with lots of X's in them. They are not very useful when we have suppressed all of the data. But we have tried drawing maps and we are trying to apply the confidentiality rules to geographical cells, so if you need three observations in the cell in order to publish a number in a table perhaps we can keep expanding the geography until we have three observations from somewhere. This is all very experimental and it is driving us crazy but it does allow you to get some interesting sub-province information.

Structure is an issue. Every OECD country is a service economy and for most of them that has been true since the 1950s. We seem to keep discovering this fact but it has been with us for a very long time. That is not just a consequence of information and communication technology development, but it is a major factor. We have laid out that infrastructure and now people are doing interesting things with it. They are developing knowledge products which are bought and sold and cannot be dropped on your foot. Have you bought any interesting financial instruments lately? Health diagnostics can be done using the internet. There are learning packages of which you could not have dreamed five years ago.

And this raises a question about practices, not just technologies. Large firms are using a set of practices to manage their knowledge about their clients and about their suppliers and about the transformation process that converts inputs into outputs, and they are doing this at a very high level in a business environment which is changing hourly. But we are not addressing those issues in official statistics. There will be an OECD book on the subject coming out in the next couple of months. But what this underscores not just the importance of linkages and the transforming effect of alliances but the importance of social sciences and humanities, of which we have spoken very little today. That is an area that we are all going to have to address.

JOHN ALIC: Let me pose a question to Fred Gault, because he brought up the practices question. My colleagues and I have done some field work in service companies in which it is apparent that they don't think of themselves as doing R&D and yet they use technology in some sense, not just infrastructural technology but technology in the sense of new and evolving knowledge to modify, alter, redesign their business practices, what the people in their organizations do on a day to day basis. They are reshaping how their organizations function and

S&T is a critical input to that but not the kind of S&T that we associate with laboratories and scientists and so on. As somebody who deals with the sources of the data, do you see a way to get a handle on that statistically?

FRED GAULT: Thank you for that question. We had an interesting problem when the banking sector claimed an enormous tax benefit for doing research and development, and a lot of that was software. So software R&D is fine. We want to work that out. The issue of organizational change and whether or not it involves R&D takes us to the edge of the Frascati definition. But so long as there is uncertainty in the undertaking, you don't necessarily know the outcome until you've got a development project that you can move with and then you think you know the outcome and you hope it works. I see no reason why that could not be classified as research and development.

As it is I come from a country where our RD-1 equivalent deals only with natural sciences and engineering, and the R&D tax benefit goes only to firms engaged in natural sciences and engineering. So all the social science R&D in industry, which is really what you are describing, is not counted, and I think that is a serious gap in our statistical system and perhaps my distinguished colleagues in NSF could discuss how it could be addressed in this statistical system. I am trying to get out of the hot seat.

CHARLES DUKE: Let me give you an example of how we use traditional R&D for development of service products. Xerox has a very multi-billion dollar business running print shops for service manuals. One of the major innovations now in process is called lead document production, which is based on the application of manufacturing technology concepts and modeling to the structure and scheduling in print shops. It doubles and triples their output. There is a classic example of how research and development of this sort – statistical analysis that first came out of MIT, is used routinely to refine the set of offerings that generate huge sums of money on a regular basis. At Xerox over half of our revenue and almost all of our profit is in services. We have the same conceptual phase and development process for our service offerings that we have for our product offerings. We have developed the facilities management services by exactly the same process that we developed a new copier or a new printer. So from our undoubtedly parochial point of view we apply very much the same discipline-oriented research to the development of new services as to the development of new tangible products, and this will become an increasingly important source of Xerox Corporation's profitability.

BILL LONG: In the manufacturing context there has always been product R&D and process R&D, and we have never had a problem calling both of them R&D. But then we get into the services, and it seems as if we have lost our mind with respect to the product/process difference. Yet the U.S. Patent Trademark Office issues patents on something called business methods or business processes. If you can get a patent on a business methods, doesn't that imply that you can do research and development to create the technology on which you got the patent? There is a lot of controversy over whether there ought to be business method patents as there is about software patent rights, but they are technologies.

JOHN JANKOWSKI: I don't want to address that question of whether Xerox should or should not be including that in its RD-1 response, but we do give guidance to specifically exclude social science R&D from the industrial research totals. We had a serious concern that companies would start including market research in their R&D totals and we felt pretty confident that we did not want that to be reported on this survey. We do indeed collect and include in our statistics social science research when it is performed by university sector or within the government sector and on occasion estimate it for not profit organizations.

JOHN ALIC: Most of what we are talking about here is not really social science R&D. There is no more social science R&D in the services than there is in manufacturing. This is industrial engineering and modeling and simulation; the analogy is factory production systems, just in time logistics management and so on.

CHARLES DUKE: My understanding is that if you guys want to get this straight, you need to go back and get your value chain straight because what you are really saying is that activity at the front end of the value chain -- how big is the market, would this product be successful if offered, etc. -- does not fit your scheme but the response to those market demands is allowable. That is my understanding of what has been said at this meeting.

FRED GAULT: That is certainly consistent with the way in which we capture the data at Statistics Canada.

PARTICIPANT: I want to address the confidentiality issue. In RADIUS we have what we call restricted but unclassified data, and only certain people can get to it. Have you in your collections from industry considered introducing this kind of concept?

PARTICIPANT: I recently retired from the National Center for Health Statistics, a federal statistical agency. What they have been doing and what the Census Bureau has been doing is setting up what they call research data centers, and these allow qualified researchers access to data that are not released for confidentiality reasons. That model is now considered fairly successful..

BILL LONG: It is my understanding that the IRI CIMS database works a little bit like that. No company that contributes data to it can go in and get access to the raw data because then they would have access to competitors' proprietary data. But almost any academic researcher could get access, subject to restrictions.

BRONWYN HALL: Our thanks to the planning committee and to all of the speakers and participants for their contributions to the discuss ion today. The workshop is adjourned.

Appendix A: Workshop Agenda

Board on Science, Technology, and Economic Policy
The National Academies

Workshop on Research and Development Data Needs

April 7, 2003

500 Fifth Street, N.W.
Washington, DC

8:30 AM   **Introduction:** Bronwyn Hall, University of California, Berkeley

8:45 AM   **Users and Uses**

There are many indications that research, development and innovation have become more critical to economic growth, corporate performance and management, and national government missions and have changed in composition, orientation, organization, location, etc.; but public data are collected in essentially the same ways and in some cases not at all. What new, prospective, or higher priority uses (e.g. national economic accounting, international comparisons, government or corporate management) will influence the demand for national R&D data and perhaps require changes in what is collected and how?

Chair: Lawrence Brown, Wharton School of Business

Speakers:
      Barbara Fraumeni, Bureau of Economic Analysis
      Andrew Wyckoff and Dominique Guellec, Organization for Economic Cooperation and Development
      David Trinkle, Office of Management and Budget
      Gregory Tassey, National Institute of Standards and Technology

10:00 AM   **Recent Developments in NSF's R&D Data Portfolio**

John Jankowksi, Science Resources Statistics Division, NSF

10:15 AM   **Composition of R&D**

We need information on the composition of R&D linked to R&D outputs (on the private side, innovation and productivity; on the public side, publications and human resources). But it is not clear that current categories (either basic/applied/development or the research field taxonomies) are useful, up-to-date, and meaningful to respondents. How can this situation be improved? Do we need better information on cross-disciplinary research?

Industry data are collected at the corporate level, obscuring the heterogeneity of activity in large, diversified companies that account for the majority of R&D and distorting the allocation across sectors. Moreover, there are many uncertainties about technical activity in the service sector. What level of detail on activities in diversified companies is it feasible to collect? Does corporate "management by the numbers" make it easier? How can we illuminate technical activity in service companies?

Chair: William Barron, Princeton University

A. Field of public sector (federal and university) research
    Michael Saltzman, Department of Energy
    Charlotte Kuh, National Research Council

B. Area of business R&D activity
    Charles Duke, Xerox Corporation
    William Long, Business Performance Research Associates, Inc.
    Ron Jarmin, Center for Economic Studies, U.S. Bureau of the Census
    Michael Gallaher, Research Triangle Institute
    John Alic, Consultant

12:00 PM   **Lunch**

12:45 PM   **R&D Collaborations**

Corporations are outsourcing more R&D and collaborating with rivals, customers, suppliers, and university scientists; but the data capture only a narrow slice of these relationships and reveal little about the magnitude of the investments, technical foci, or duration. Are private data sources adequate over the long term? What do firms trade? What relationships do we want to capture on an ongoing basis and how should we do that?

Chair: Don Siegel, Rensselaer Polytechnic Institute

Speakers:
Nick Vonortas, George Washington University
Josh Lerner, Harvard Business School

2:00 PM   **Break**

2:15 PM   **Location of R&D Activity**

Where R&D is performed is important to public officials at all levels. Domestically, the clustering of technology-based industry and associated infrastructure (including research universities) is viewed as critical to growth, and the focus of public economic development support is frequently regional or local; but data are only national, state, and institutional. If finer-grained industrial data were collected, would it be possible to protect the confidentiality of respondents? Cross-national investment is increasing rapidly, and the current survey collects some data, but there are gaps. How can they be overcome?

Chair: Patrick Windham, Windham Consulting

A. Local
    Rob Atkinson, Progressive Policy Institute
B. International
    Walter Kuemmerle, Harvard Business School
    Obie Whichard and Ned Howenstine, Bureau of Economic Analysis
    Robert McGuckin, The Conference Board

4:00 PM   **Concluding Observations**

Al Johnson, Corning and the Industrial Research Institute
Bronwyn Hall, University of California, Berkeley
Fred Gault, Statistics Canada

5:00 PM   **Adjourn**

Appendix B: List of Participants

John Alic
Consultant

Peder Andersen
U.S. International Trade Commission

Rob Atkinson
Progressive Policy Institute

Barbara Bailar
Consultant

William Barron
Princeton University

Rich Bennof
National Science Foundation

Mark Boroush
Department of Commerce

Lewis Branscomb
Harvard University

Lawrence Brown
Wharton School of Business

Alphonse Buccino
Contemporary Communications, Inc.

Lynda Carlson
National Science Foundation

Leslie Christovich
National Science Foundation

Stacey Cole
Bureau of the Census

K. C. Das
Office of Science and Technology

Michael Davey
Congressional Research Service

Charles Duke
Xerox Research & Technology

Elisa Eiseman
RAND

Chad Evans
Council on Competitiveness

Donna Fossum
RAND

Mary Frase
National Science Foundation

Barbara Fraumeni
Bureau of Economic Analysis

Michael Gallaher
Research Triangle Institute

Fred Gault
Statistics Canada

Gurmukh Gill
Economics and Statistics Administration

Katherine Gill
ORC

David Gromos
Bureau of the Census

Dominique Guellec
Organization for Economic Cooperation and Development

Randall Haley
EPSCoR Foundation

Bronwyn Hall
University of California, Berkeley

Jake Haselswerdt
Lewis-Burke Associates

Robert Hershey
Robert L. Hershey, P.E.

Christopher Hill
George Mason University

Derek Hill
National Science Foundation

Ron Hira
Columbia University

Ned Howenstine
Bureau of Economic Analysis

Paul Hsen
Bureau of the Census

Cassandra Ingram
Economics and Statistics Administration

Yongsuk Jang
George Washington University

John Jankowski
National Science Foundation

Ron Jarmin
U.S. Census Bureau

Albert Johnson
Corning Incorporated

Sang-Seon Kim
Korean Embassy

Walter Kuemmerle
Harvard Business School

Charlotte Kuh
The National Academies

Catherine Langrehr
Association of American Universities

Carolyn Lee
University of California, San Diego

Rolf Lehming
National Science Foundation

Michelle Lennihan
Council on Competitiveness

Josh Lerner
Harvard Business School

William Long
Business Performance Research Associates, Inc.

Marge Machen
National Science Foundation

Sue Majewski
Department of Justice

Maggie Marcum
USG

Merrilea Mayo
The National Academies

Robert McGuckin
The Conference Board

Don McMaster
ORC

Bruce McWilliams
George Washington University

Ron Meeks
National Science Foundation

Reese Meisinger
ASME International

Stephen A. Merrill
The National Academies

Mary Ellen Mogee
Mogee Research & Analysis

Kim Moore
Bureau of the Census

Yvette Moore
Bureau of the Census

Mark Morgan
ORC

Francisco Moris
National Science Foundation

Russell Moy
The National Academies

Jerri Mulrow
National Science Foundation

David Napier
Aerospace Industries Association

Sumiye Okubo
Bureau of Economic Analysis

Tom Plewes
The National Academies

David Radzanowski
Office of Management and Budget

Alan Rapoport
National Science Foundation

Proctor Reid
National Academy of Engineering

Michael Reischman
NASA

Sally Rood
Federal Demonstration Partnership

Michael Saltzman
Department of Energy

John Sargent
Office of Technology Policy, Technology Administration

Joshua Sarnoff
Washington College of Law, American University

Harold Schmitz
Mars, Incorporated

Craig Schultz
The National Academies

Brandon Shackelford
National Science Foundation

Don Siegel
Rensselaer Polytechnic Institute

Julius Smith
Bureau of the Census

Tim Smith
WESTAT

Byongho Son
George Washington University

Gregory Tassey
National Institute of Standards and Technology

Rebecca Trager
Research

David Trinkle
Office of Management and Budget

Richard Turman
Association of American Universities

Patrick von Bargen
Council on Competitiveness

Nick Vonortas
George Washington University

Kathleen Walsh
Henry L. Stimson Center

Philip Webre
Congressional Budget Office

Amanda Welch
Council on Competitiveness

Charles Wessner
The National Academies

Obie Whichard
Bureau of Economic Analysis

Andy White
The National Academies

Roger Whiteley
Industrial Research Institute

Yvette White
The National Academies

Patrick Windham
Windham Consulting

Ray Wolfe
National Science Foundation

Andrew Wyckoff
Organization for Economic Cooperation and Development